Fuels and Petrochemicals Division 2019

Held at the 2019 AIChE Spring Meeting and 15th Global Congress on Process Safety

New Orleans, Louisiana, USA
31 March - 4 April 2019

ISBN: 978-1-5108-8702-2

Printed from e-media with permission by:

Curran Associates, Inc.
57 Morehouse Lane
Red Hook, NY 12571

CURRAN ASSOCIATES INC. proceedings.com

Some format issues inherent in the e-media version may also appear in this print version.

Printed by Curran Associates, Inc. (2019)

For permission requests, please contact AIChE
at the address below.

AIChE
120 Wall Street, FL 23
New York, NY 10005-4020

Phone: (800) 242-4363
Fax: (203) 775-5177

www.aiche.org

Additional copies of this publication are available from:

Curran Associates, Inc.
57 Morehouse Lane
Red Hook, NY 12571 USA
Phone: 845-758-0400
Fax: 845-758-2633
Email: curran@proceedings.com
Web: www.proceedings.com

TABLE OF CONTENTS

551622 Techno-Economics of Catalytic Conversion of Ethanol to Chemical Grade n-Butanol and 1-Hexene

A techno-economic analysis of an ethanol to n-butanol process was developed to investigate the feasibility of making chemical grade products that have a premium price relative to transportation fuels. To remove the uncertainties associated with the various processes for making ethanol, this analysis assumed that the ethanol feed stock for the process would be purchased as a commodity chemical at market prices and upgraded to n-butanol which has a significantly higher value as a chemical. Mixed oxide based catalyst developed for this process has demonstrated a 61% one pass ethanol conversion to n-butanol, with some 6- and 8-carbon alcohols (linear and branched). A conceptual process to upgrade the n-hexanol fraction of the co-products to produce chemical grade 1-hexene was developed to improve the minimum butanol selling price (MBSP) – the price that gives a net present value of zero over the life of the project after accounting for all costs and revenues, including a 10% internal rate of return, using a discounted cash flow method. The results of this study will be presented including economic sensitivities to the ethanol feed stock cost, 1-hexene market value, conversion and selectivity to products and co-products, plus other financial assumptions.

551621 Estimating Reid Vapor Pressure for Mixtures of Oxygenates Blended in a CA-Rbob

The Reid Vapor Pressure (RVP), the vapor pressure measured at 37.8 °C, is a critical quality attribute for gasoline set by EPA regulations to reduce the evaporative loss of gasoline to the environment. Various components of gasoline can impact air quality through the formation of smog and increased ozone formation. When blended with gasoline, some organic oxygenates such as ethanol, disproportionately increase the RVP beyond its own pure component vapor pressure because of non-ideal mixing; some oxygenates can decrease the RVP when blended into an E10 gasoline-ethanol mixture, as well. The deviations from ideal mixing characteristics are dependent on the formulation of the gasoline and the molecular interactions between the many compounds present in gasoline.

The Co-Optimization of Fuels and Engines initiative supported by the U.S. Department of Energy (Co-Optima) is investigating the concurrent optimization of fuels and engines to achieve higher efficiency with lower environmental impact. A variety of new oxygenated blend stocks have been proposed as candidates for further study. The RVP of the oxygenates/gasoline mixtures will be a critical characteristic of each candidate that will help determine the optimum blend level, and to some extent its market value. The ability to predict the RVP of new mixtures facilitates the evaluation process by eliminating the need to experimentally measure every possible candidate and allows researchers to identify possible mixtures that have beneficial (decreased) RVP changes when blended in the right proportions. Promising candidates can then be validated experimentally.

A California Air Resources compliant Reformulated Blend stock for Oxygenate Blending with 10 volume percent ethanol added (e.g. - E10 CA-RBOB) was blended with various oxygenates including higher alcohols, ketones, acetates, and acetals; the RVP was measured by a commercial laboratory using the ASTM D6378 test method. An ASPEN Plus model was used to emulate the ASTM method, and the RVP values were estimated for each mixture at 10, 20 and 30 volume percent oxygenate in an E10 CA-RBOB. The physical property models in ASPEN Plus (version 10) that were used for the RVP estimates were the Cubic Plus Association equation of state (CPA), and the Non-Random Two Liquids (NRTL), Wilson, UNIFAC and UNIF-DMD activity coefficient models. Compared to experimental measurements, the RVP estimates made using the CPA, NRTL and Wilson equations had average relative errors over all 57 samples of -1.5%, -6.4%, and -9.6%, respectively, with worse case errors of -11%, -19%, and -22%. The UNIFAC and UNIF-DMD models had relative errors of -2.6% and -0.6% over the 45 samples for which there were default parameters in the ASPEN Plus databank; four of the chemicals used in the study had missing parameters for the UNIFAC equations, so their average errors cannot be compared directly with the models for which all components were included.

The results of this study were based solely on physical property data and binary interaction parameters that come with the ASPEN Plus software package with no attempt to provide new interaction parameters based on new data regressions. Better binary interaction parameters, where applicable, could potentially improve model estimates.

551544 Quantifying the Impact of Catalyst Architecture on Yield and Activity Lifetime through Mesoscale Modeling and Simulation

Effective heterogeneous catalytic processes require precise coordination of physiochemical phenomena that occur over orders of magnitude in length and time scales. The majority of computational studies of such systems focus on the extrema: detailed investigations of reaction energetics and mechanisms at the atomic scale; or process-scale simulations of catalytic reactor systems. However, many important processes occur at intermediate scales wherein transport of reactants and products within catalyst particles are tightly coupled to the rates of desirable and undesirable chemical reactions. Tunable catalyst design parameters such as particle size, shape, and porosity can have a large impact in the effective yield, selectivity, and activity lifetime delivered by a catalyst in given reactor system. We have been working to develop "mesoscale" modeling approaches to understand and quantify these effects and thereby provide actionable recommendations to catalyst designers. In this presentation, I will overview two examples of mesoscale simulations in catalytic systems used for the transformation of bio-derived intermediates. First, I will present a study of catalytic fast pyrolysis wherein ex-situ vapor-phase upgrading of pyrolysis vapors is performed using a zeolite catalyst. Next, I will describe packed-bed catalyst system that utilizes a structured mesoporous silica support (SBA16) to convert ethanol to butadiene. In both cases I will present methods to develop suitable hierarchical transport models coupled to chemical reaction kinetics and provide insight for improving the performance of the system based on modification of architectural features of the catalyst particles.

551938 Exploring Options for Cryogenic NGL Recovery from Natural Gas with Emphasis on Process Optimization

Recent boom in natural gas production and utilization creates new opportunities for natural gas liquids (NGL) recovery. As demand for Ethane fluctuates with the price, the need for optimization is vital. The same is true about Propane and Butanes products (light petroleum gas or LPG).

There are several thermodynamic cycles for NGL production. Some are suitable for both ethane recovery and rejection, others only for ethane rejection and high propane recovery. The cycles vary in complexity, cost and efficiency. There is also a need for flexibility between ethane rejection and recovery, low cost solutions (capital investment and operating cost), and for retrofits.

The following paper provides an overview of NGL cycles with emphasis on differences in complexity, flexibility, process optimization and process screening by case studies. This will look at different NGL cycles with changing feeds to these processes.

550240 Molecular Modelling of Co-Processing Biomass Pyrolysis Oil with Vacuum Gasoil in an Oil Refinery Fluid Catalytic Cracking Unit

Integration of renewable resources in petroleum refining has been a research focus for several years. It is becoming a priority for several developed countries, especially in Europe and North America, to increase the share of renewable energy sources in their transportation fuel system. For example, EU's members countries aim to increase renewables contribution from the current value of 7.1% to 10% by 2020 [1] .

Several works try to find optimum integration points within transportation fuel system [2], while other researches focus on assessing the viability of different renewable sources. Recent works have been focusing on the integration of second generation biomass feedstock in petroleum refining. This is mainly due to the fact that second generation biofuels do not create competition with food industry and satisfy the criteria published in Article 17 [3] .Two types of processes are suggested, namely gasification and pyrolysis. This research is concerned with the second alternative since it allows the assessment of several integration options within conventional petroleum refinery for the bio-liquid asset produced.

Bio oil produced from pyrolysis processes is considered as a candidate for co-processing with heavy petroleum fractions in refinery catalytic conversion processes [4], such as catalytic cracking and hydroprocessing units. Careful assessment of the effects of introducing bio oil to refinery processes is required.

In this work, a novel methodology is developed to investigate the effect of introducing bio oil to the feed of the fluid catalytic cracking units on product yields and properties especially gasoline and diesel. The proposed method includes detailed, molecular-level characterisation of feed and product streams based on Molecular Type and Homologous Series matrix (MTHS) approach, as well as detailed reaction network and kinetic models which take into account the complex interactions between the various molecular attributes. Kinetic parameters is tuned through an optimisation problem in which the objective function is to minimise the difference between measured and predicted products yields at various conversion levels. The capability of the proposed methodology is demonstrated using a case study for the characterisation of vacuum gas oil (VGO) and fast pyrolysis oil (FPO) blend which is used as a feed for fluid catalytic cracking process (FCC) [5]. There are 328 reactions that are used for the synthesis of the reaction network. The computational results demonstrate that an overall good agreement between measured and predicted yields is obtained using the developed kinetic model for VGO: FPO blending ratio, C/O ratio and reaction temperature of 95:5, 5 and 530°C, respectively [6]. PONA composition in each layer of product stream (e.g. gasoline, diesel, gasoil, etc.), as well as oxygen compounds compositions and oxygen content in each product fraction is predicted using the developed models.

This creates potentials for rigorous optimisation of process parameters for refinery process maximisation or better products' quality control. The methodology developed in this work can easily be extended for the modelling of other refinery processes. This will help in rigorously evaluating different integration options of biomass pyrolysis oil into conventional refinery, and will help into integrating the developed reaction models into dynamic models for refinery's hydrogen network in order to optimise hydrogen consumption. Finally, the success of this approach can revolutionise refinery processes' design methodologies and optimise the integration of renewable resources, such as biomass, into transportation fuels system.

KEYWORDS: Catalytic cracking, Fast pyrolysis oil (FPO), biomass, refinery, MTHS matrix, Vacuum gas oil (VGO)

[1] The European Commission, “Renewable energy statistics,” 2015. [Online]. Available: http://ec.europa.eu/eurostat/statistics-explained/index.php/Renewable_energy_statistics. [Accessed: 01-Jun-2017].

[2] S. Jones *et al.*, “Process Design and Economics for the Conversion of Lignocellulosic Biomass to Hydrocarbon Fuels Fast Pyrolysis and Hydrotreating,” *PNNL Rep.*, vol. PNNL-23053, p. 97, 2013.

[3] European Commission, “Sustainability criteria,” 2011. [Online]. Available: https://ec.europa.eu/energy/en/topics/renewable-energy/biofuels/sustainability-criteria. [Accessed: 25-May-2017].

[4] D. C. Elliott, “19 - Production of biofuels via bio-oil upgrading and refining BT - Handbook of Biofuels Production (Second Edition),” Woodhead Publishing, 2016, pp. 595–613.

[5] D. V Naik *et al.*, “Catalytic cracking of jatropha-derived fast pyrolysis oils with VGO and their NMR characterization,” *RSC Adv.*, vol. 5, no. 1, pp. 398–409, 2015.

[6] D. V Naik, V. Karthik, V. Kumar, B. Prasad, and M. O. Garg, “Kinetic modeling for catalytic cracking of pyrolysis oils with VGO in a FCC unit,” *Chem. Eng. Sci.*, vol. 170, pp. 790–798, 2017.

544202 The Process Chemistry of Lipid Hydrodeoxygenation

Commercial production of Renewable Diesel (RD) from fats and oils began in 2007, and has grown to about 1.5 billion gal/y in just over ten years. The growth is driven not only by national and local regulatory regimes promoting low carbon fuels, but also quality. With a composition made entirely of n-paraffins and methyl-branched iso-paraffins, RD has all the desirable molecules of conventional hydrocarbon diesel without the undesirable polyaromatics present in petroleum middle distillates. As a high cetane, clean burning fuel with good low temperature properties, RD is not only a drop-in biofuel, but a premium diesel.

The key step in production of RD is hydrodeoxygenation (HDO). This is the reaction that converts fatty acid/glycerides into paraffins. Although the HDO paraffin product is typically hydroisomerized for RD production, it is also an excellent steam cracker feed for olefins. Depending on feedstock, HDO paraffins may also be used as phase change material or in other specialty paraffin applications.

Multiple HDO reaction pathways are possible, each influenced by type of catalyst and reactor conditions (and to some extent, the lipid feed itself). In this paper, we will review the various reaction pathways and discuss the corresponding mechanisms. Two reaction pathways in particular, one involving an aldhehyde intermediate and the other based on a carboxylic acid intermediate, are presented and discussed in terms of observations from commercial and pilot plant reactor systems.

Role of catalyst type in promoting various reaction pathways is discussed, including type of hydrogenation metals (active sites) and support porosity. HDO catalyst deactivation mechanisms are reviewed based on actual commercial feeds and fixed-bed reactor performance data. Side-reactions and their kinetics relative to desirable HDO pathways are also reviewed.

Additionally, "hydrogen free" approaches to deoxygenation of lipids to hydrocarbons are discussed. Recommendations for future research in view of RD production needs are also presented.

19 AIChE Spring Meeting
15TH GLOBAL CONGRESS ON PROCESS SAFETY
NEW ORLEANS • 2019
March 31 – April 4, 2019
Hilton New Orleans Riverside
Safer and Improved Process Design of LNG Plants Using Dynamic Simulation
Presenters:
Simon Leyland, Process Systems Enterprise
Matthew Okasinski, Air Products

LNG production with multiple compressor strings

Presenter:
Matthew Okasinski, Air Products

Good Afternoon,

In recent years, some baseload LNG projects have chosen to use parallel refrigeration strings to improve turndown efficiency, compressor aerodynamics, and on-stream availability. However, this configuration introduces operating consequences such as how to smoothly transition between 2 string operation and single string operation without causing a sympathetic trip of the surviving refrigeration string. In addition, the transition between 1 string to 2 string operation also needs careful consideration during the detailed design phase of the project, however due to time constraints, I'll only discuss loss of a string.

Losing one of two strings represents a 50% loss of power into the process. Because LNG production is proportional to the power consumed by the refrigeration system, the new steady-state production after the transition will be approximately 50% of design production. As the number of parallel strings (n) increases, there is an increased equipment cost due to duplicating the refrigeration string (n-1) times but the production lost by the trip of one string decreases by approximately maximum production / (n). As two parallel refrigeration strings is the most common configuration and causes the largest change in available refrigeration, the example will focus on a liquefaction process with two parallel mixed refrigerant strings. However, the learnings also apply to configurations with more than 2 string.

LNG production with multiple compressor strings

- Parallel strings
  - Improve turndown efficiency
  - Compressor aerodynamics
  - On-stream availability
  - Modify operating procedures for transition between number of strings online
- Power into liquefaction process
  - Proportional to production attainable
  - Loss of 1 of n strings causes a production decrease ≈ 1/n
  - Parallel string economics

Good Afternoon,

In recent years, some baseload LNG projects have chosen to use parallel refrigeration strings to improve turndown efficiency, compressor aerodynamics, and on-stream availability. However, this configuration introduces operating consequences such as how to smoothly transition between 2 string operation and single string operation without causing a sympathetic trip of the surviving refrigeration string. In addition, the transition between 1 string to 2 string operation also needs careful consideration during the detailed design phase of the project, however due to time constraints, I'll only discuss loss of a string.

Losing one of two strings represents a 50% loss of power into the process. Because LNG production is proportional to the power consumed by the refrigeration system, the new steady-state production after the transition will be approximately 50% of design production. As the number of parallel strings (n) increases, there is an increased equipment cost due to duplicating the refrigeration string (n-1) times but the production lost by the trip of one string decreases by approximately maximum production / (n). As two parallel refrigeration strings is the most common configuration and causes the largest change in available refrigeration, the example will focus on a liquefaction process with two parallel mixed refrigerant strings. However, the learnings also apply to configurations with more than 2 string.

AP-C3MR™ Schematic

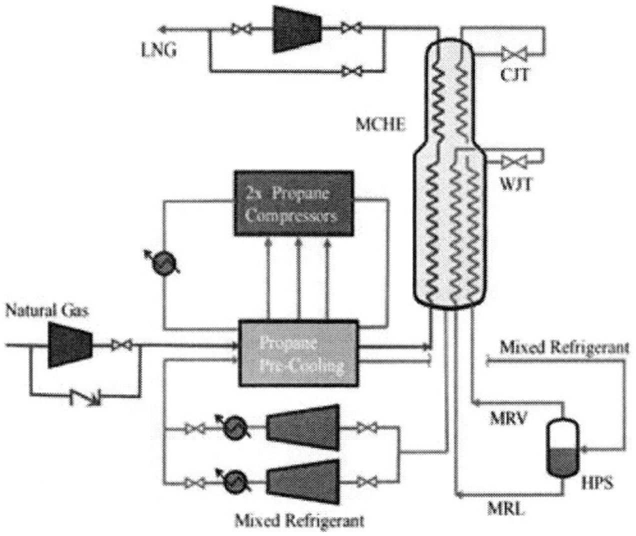

The AP-C3MR process is shown schematically here. It is a propane – mixed refrigerant (AP-C3MR™) process with two parallel mixed refrigerant (MR) and propane (C3) compressor strings. For simplicity, the two propane compressor strings are shown as a single box. Dry, clean natural gas is compressed to high pressure and then cooled against propane prior to entering the Main Cryogenic Heat Exchanger (MCHE). Within the MCHE, the feed gas is further cooled, liquefied, and subcooled against a mixed refrigerant containing nitrogen and light hydrocarbons. Upon exiting the MCHE, the LNG stream is reduced to nearly atmospheric pressure across, for example, a hydraulic turbine, valve or some other similar device. The mixed refrigerant (MR), carrying the heat removed from the tube circuits, exits the MCHE shell, is compressed to high pressure, and cooled against propane. After phase separation, the mixed refrigerant liquid (MRL) and vapor (MRV) are further cooled and liquefied within the MCHE tubes. Finally, the MRV and MRL are reduced in pressure across the warm and cold Joule Thompson valves (WJT and CJT respectively) to provide refrigeration for the MCHE.

This entire model was built into a dynamic simulation package and contains about 16,000 algebraic and differential equations. The model incorporates actual volumes, valve sizes and compressor curves among other process data.

Loss of a Refrigeration String

Disturbance: loss of refrigeration string

Avoid process trips
- Low suction temperature
- Warm LNG temperature
- Low speed on surviving GT

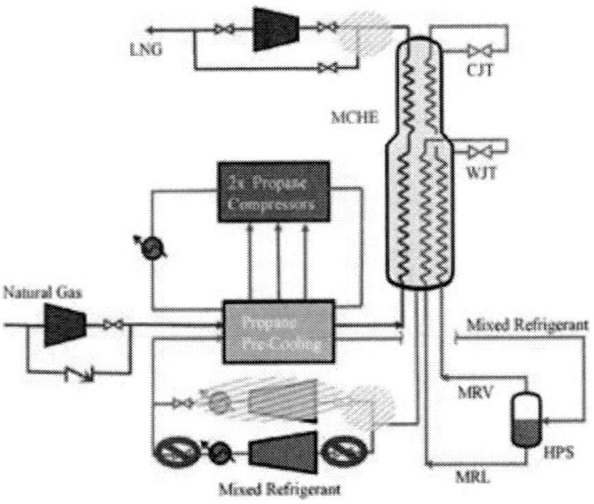

In this example, a trip signal causes the discharge isolation valve of one of the refrigeration strings to close. In response the control scheme will close the string's suction side valve to isolate the train from the process. This is a major disturbance to the liquefaction process and dynamic simulation helps us understand the impact of this disturbance as we seek to transition to another operating point while avoiding a trip of the liquefaction process.

Two events that will cause a trip of the liquefaction unit are a low MR suction temperature and high LNG outlet temperature. A third event, a sympathetic trip of the surviving compressor string would be triggered if the surviving gas turbine slowed down under increased load and tripped on low speed.

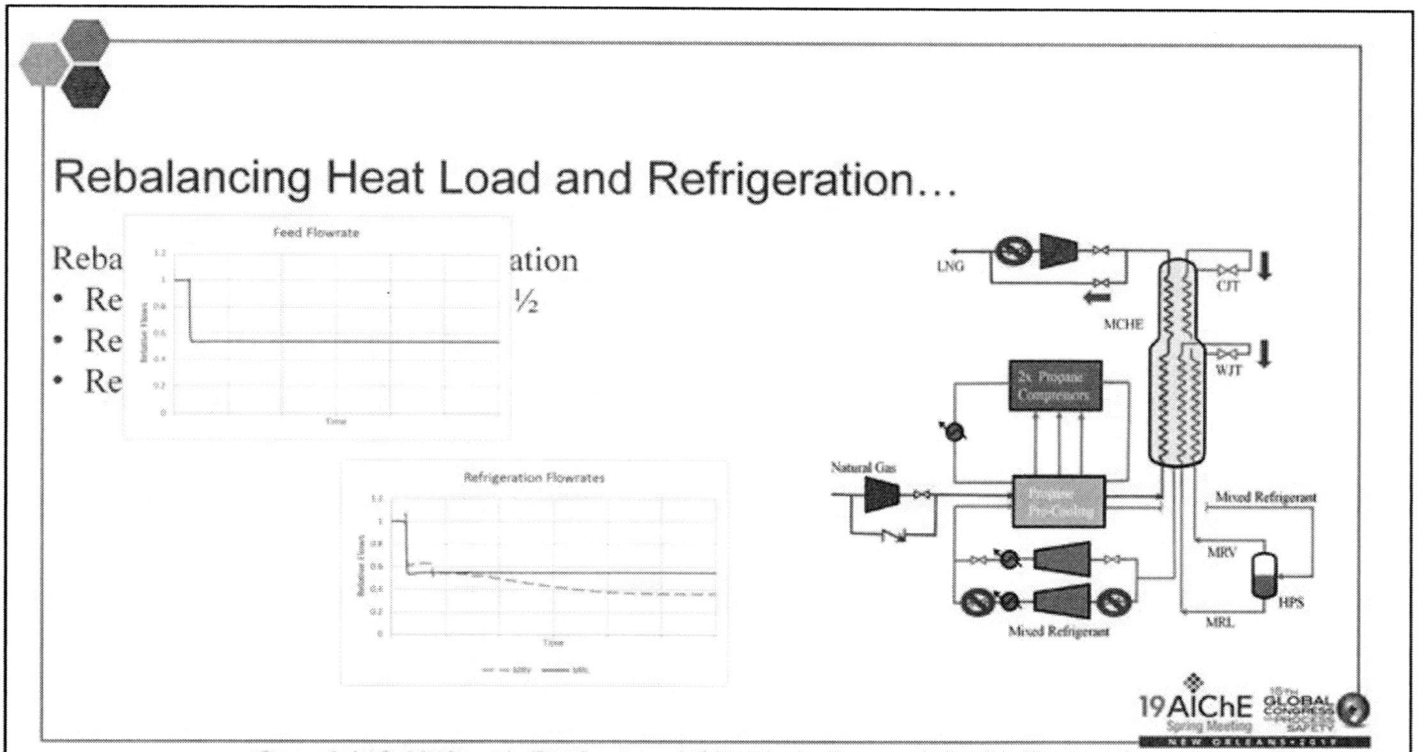

As half of the refrigeration capacity has been lost upon the string trip, the heat load and refrigeration need to be rebalanced immediately to prevent a temperature-based trip. Simultaneously the interlocks trip the LNG hydraulic turbine to protect the machine, open the LNG bypass valve a predetermined amount, and partially close the WJT and CJT a predetermined amount. Multiple dynamic simulations are used to determine these valve openings after the trip, which can then be fine tuned in the field, based on plant performance.

Here are the graphs of the Feed and refrigeration flowrates. Notice that the MRV flowrate was reduced beyond 50%. At reduced flowrates, there is more heat transfer area available so less refrigerant is needed. In addition, the MRV becomes lighter as heavier MR components accumulate in the High Pressure Separator so less MRV flow is needed.

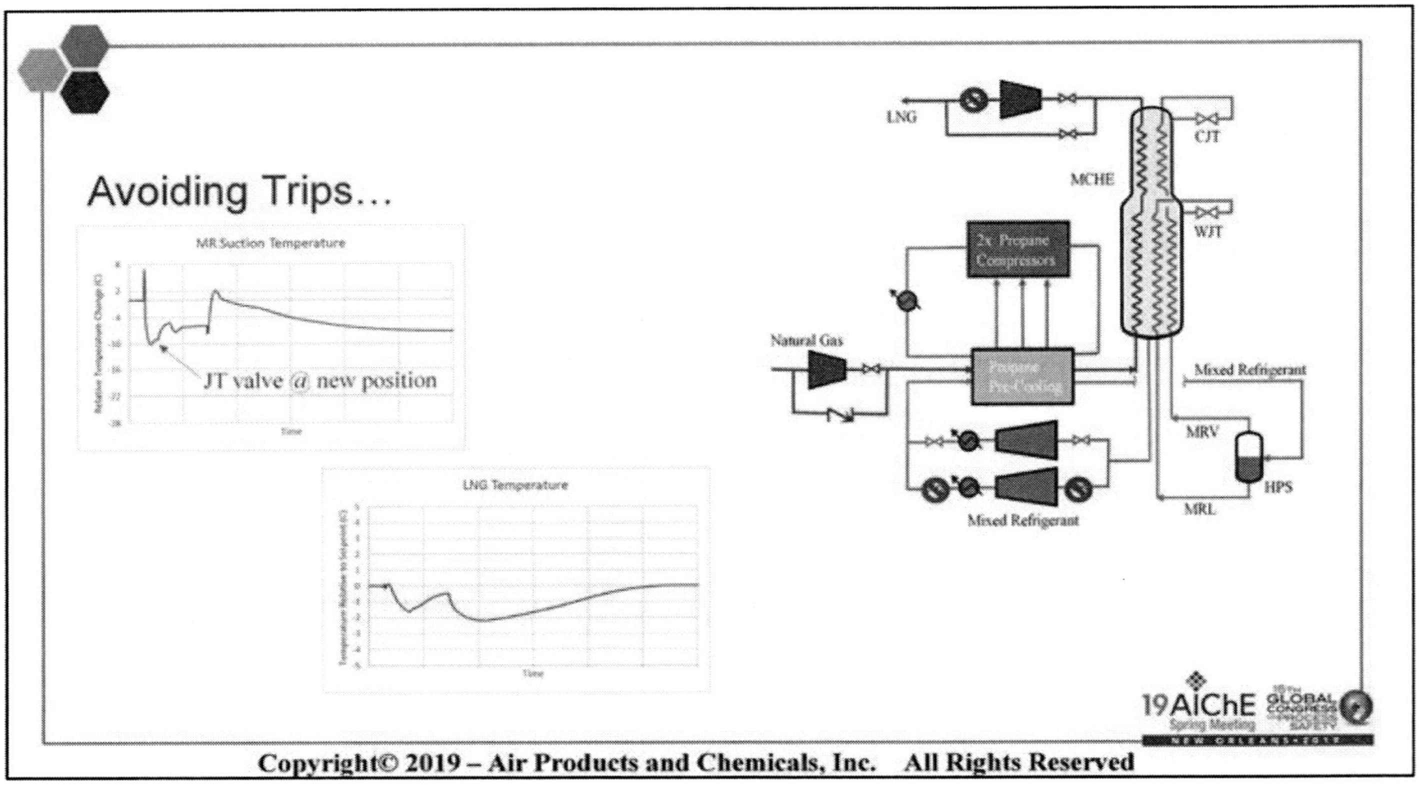

The suction temperature initially spikes warmer due to the tripped compressor antisurge valve opening and the finite time needed to close the suction isolation valve on the tripped string. The suction temperature then becomes colder as the LNG flow reduces and refrigeration accumulates in the MCHE shellside, due to only one operating compressor string. Once the LNG and WJT valves get near their pre-determined openings, the temperature begins to stabilize.

Throughout this disturbance, the LNG outlet temperature stayed within a few degrees Centigrade of its pre-trip value. Note that the LNG stayed colder throughout the transition to a lower production. With one compressor string offline, there is a reduced need for fuel gas, so the pre-determined positions of the refrigeration valves were chosen to yield a colder LNG.

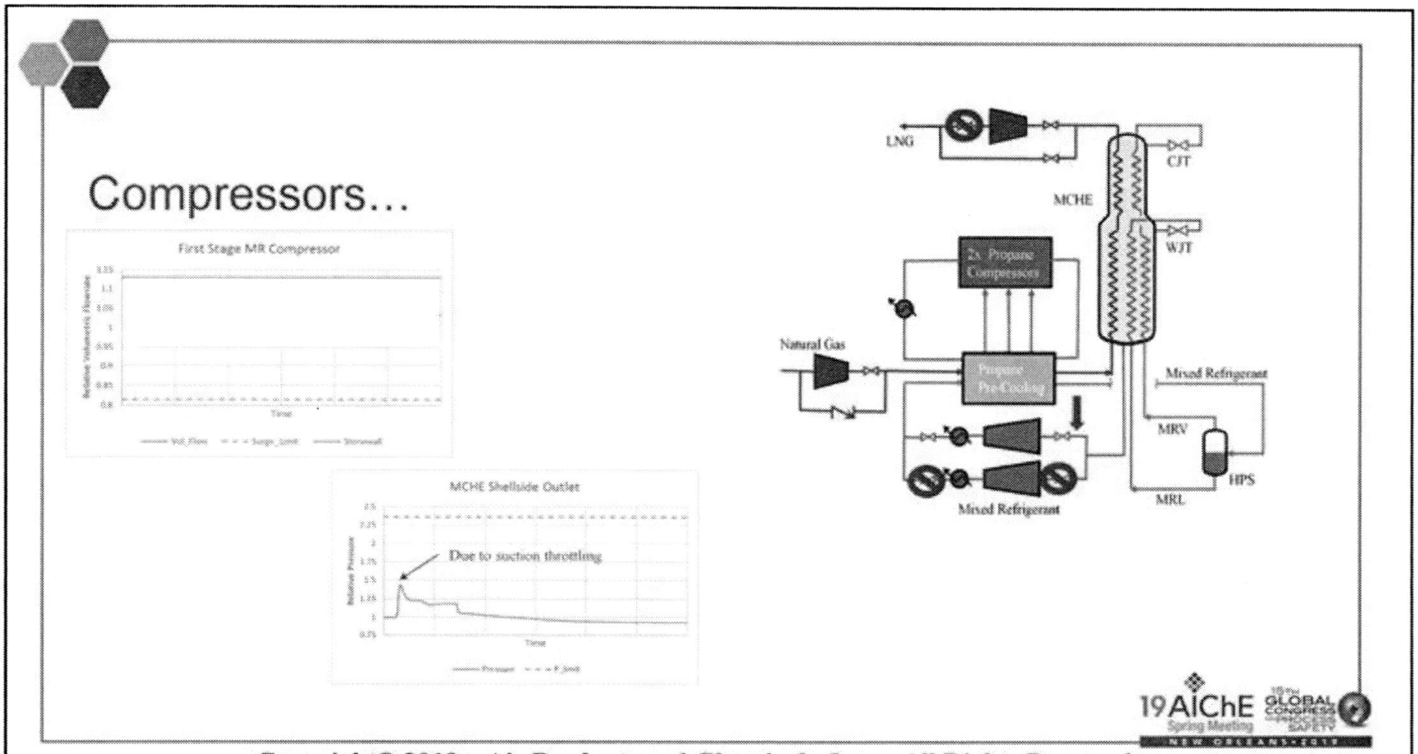

However, simply balancing heat load and refrigeration is not sufficient to ensure a transition to a reduced production operating point. When the compressor string is suddenly isolated, the flow that was going into 2 strings can now only flow through 1 string. This pushes the operating point of the surviving string's LP compressor towards stonewall, increasing the power demand from its turbine. Once the turbine reaches its maximum available power, it will slow down. A gas turbine has limited speed range and could trip on low speed – this is known as a sympathetic trip.

Through use of dynamic simulation, a methodology is developed to transition from 2 string to 1 string operation without a sympathetic trip. The goal is to keep the compressor within its normal operating range between surge and stonewall. This is accomplished by quickly *reducing (click)* the flow into the surviving string's first stage compressor suction by throttling the inlet valve. This will temporarily store excess refrigeration in the MCHE shell, raising its pressure. As with the WJT and CJT valves, many dynamic simulations are used to determine the speed and degree of throttling to keep the surviving string operating within its normal range. The suction throttle valve is fully opened once the process has stabilized.

This response strategy shows the value of dynamic simulation. One's first thought is that after losing 50% of the MR compression capacity, the best strategy is to use the remaining compressor to compress as much flow as possible. However, this approach is expected to cause a sympathetic trip. Rather, the correct action is to keep the flow to the compressor within an acceptable operating range, preventing overloading the gas turbine, and then slowly moving the extra MR from the MCHE and storing it in the High Pressure Separator.

This is just one example of dynamic simulations that were performed for a currently operating baseload plant in addition to several LNG plants currently under construction, all containing parallel compressor strings. Initial feedback from the plant shows that the model accurately predicts plant behavior.

Now I'd like to introduce my co-author ??? from Process System Enterprise will present the remainder of our paper.

Fire Survivability of Blowdown Segments

Presenter:
Simon Leyland, Process Systems Enterprise

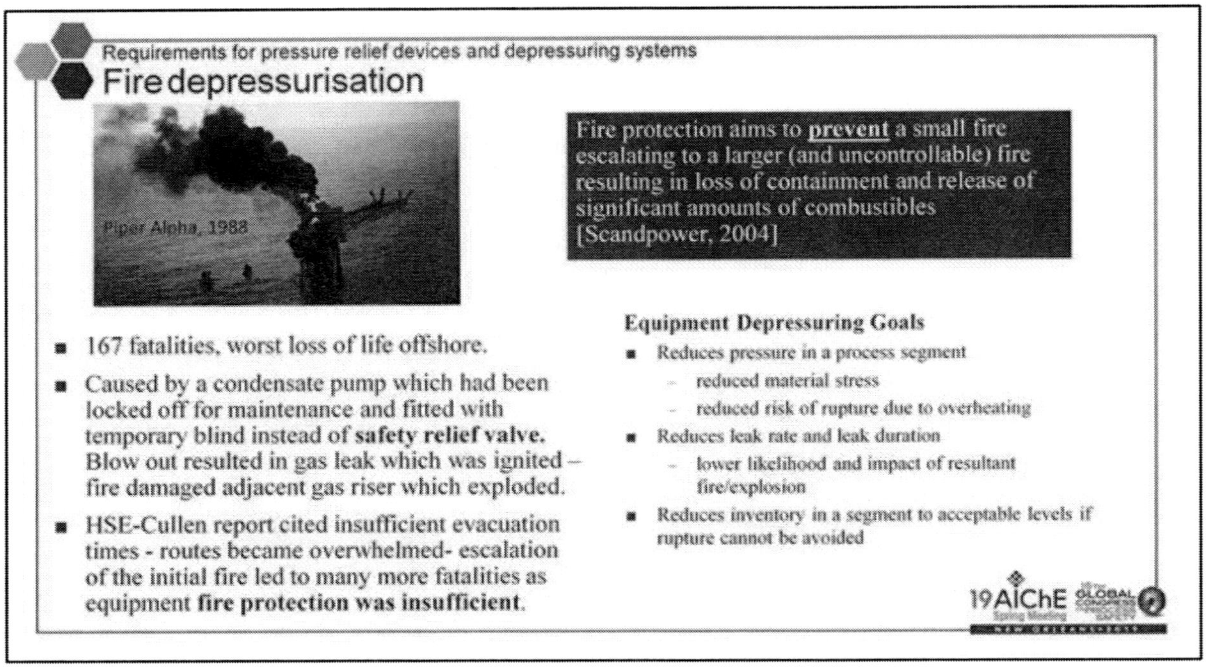

Pressure relief and blowdown are the last layers of defence available in a plant in the event of an emergency for example fire. Getting the design incorrect can lead to catastrophic consequences and Pipe Alpha incident reminds us of the consequneces of poor design

The purposes of fire protection facilities is to prevent the escalation of local fires in to large uncontrollable fires. Depressurisation of blowdown systems help

- reduce the pressure of the segment and this reduces the pressure stresses on the system,
- Safe disposal of HC inventory and reduces the risk of leakage
- Reduces inventory in a segment to acceptable levels if rupture cannot be avoided

In this presentation, I will briefly talk about how the design guidelines have changed recently and how advanced processing modelling can help arrive at safe designs

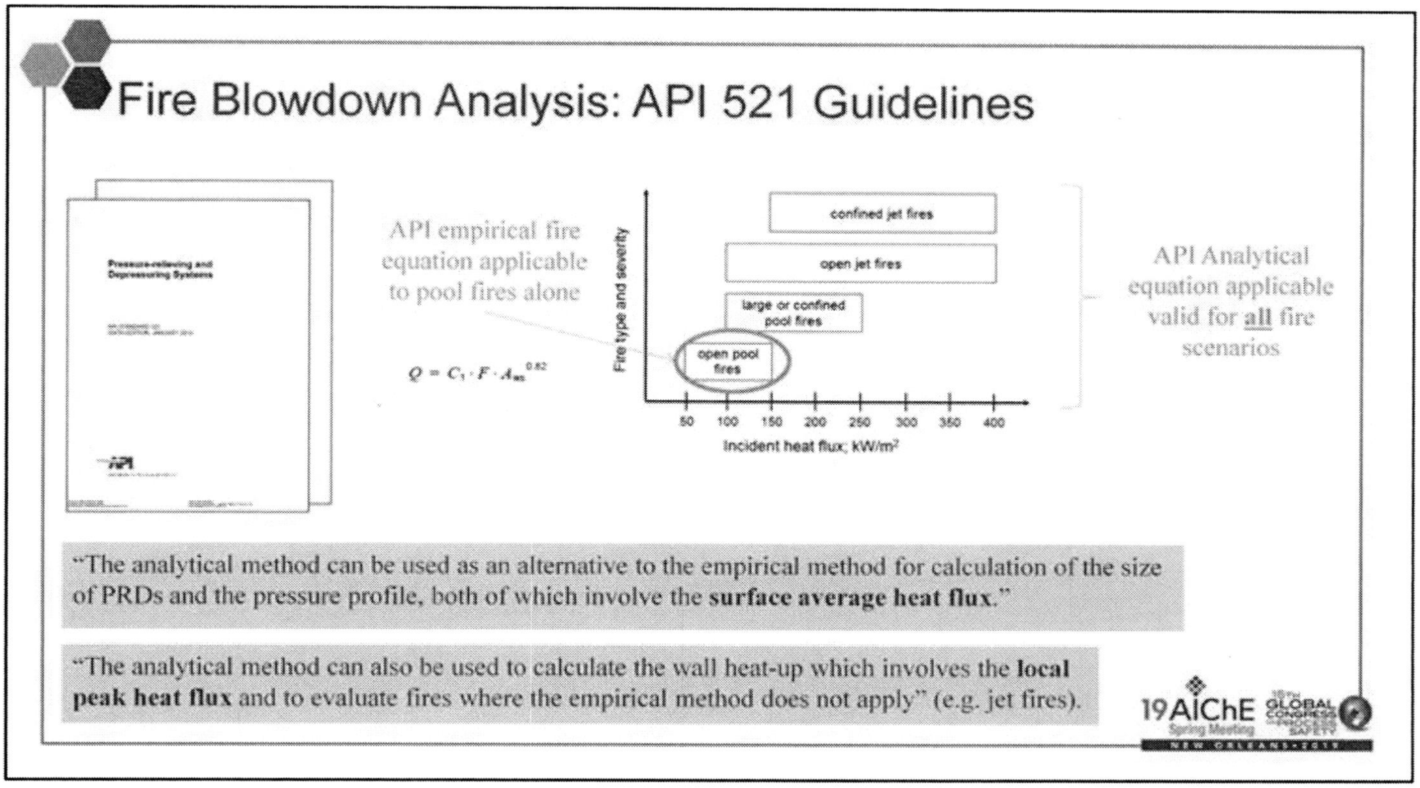

The latest version of API 521 was issued in 2014. Prior to this, API referred to an empircal equation to represent the fire heat input in to a blowdown segment to determine depressurisation rates. The application of this equation is limited to open pool fire scenarios and no clear guidelines was available for the jet fire conditions. In the current version of the guidelines, this issue has been addressed and API now provides an alternative analytical methodology for pool and jet fire blowdown scenarios. This methodology is also applicable for design of pressure relief devices.

In the event of fire, only small areas of the equipment are exposed to the peak heat fluxes while majority of the segment is exposed to lower thermal fluxes. API recommends using the peak heat fluxes to determine the wall heat up rate while the average fluxes are used to determine the pressure increase within the processe segment

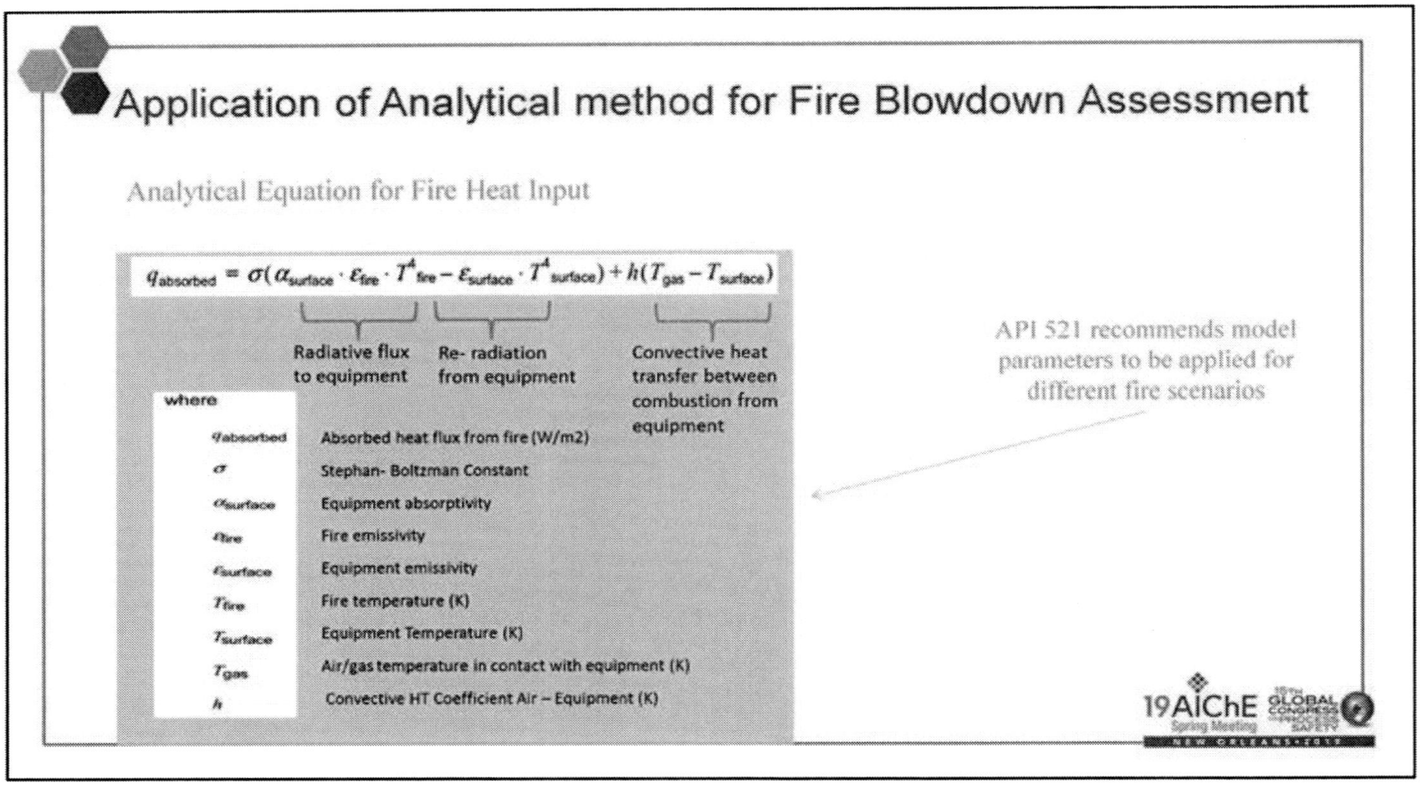

What is required to carry out a fire survivability assessment that is complaint with the API analytical methodology?

1) A tool that calculates the fire heat input based on the analytical methodology
2) The fire heat input equations is based on fundamentals and contains both the convective and radiative heat exchange between the flame and the pipe and drum metal walls
3) There are a number of model parameters involved but API 521 recommends the values to be used for the different fire scenarios

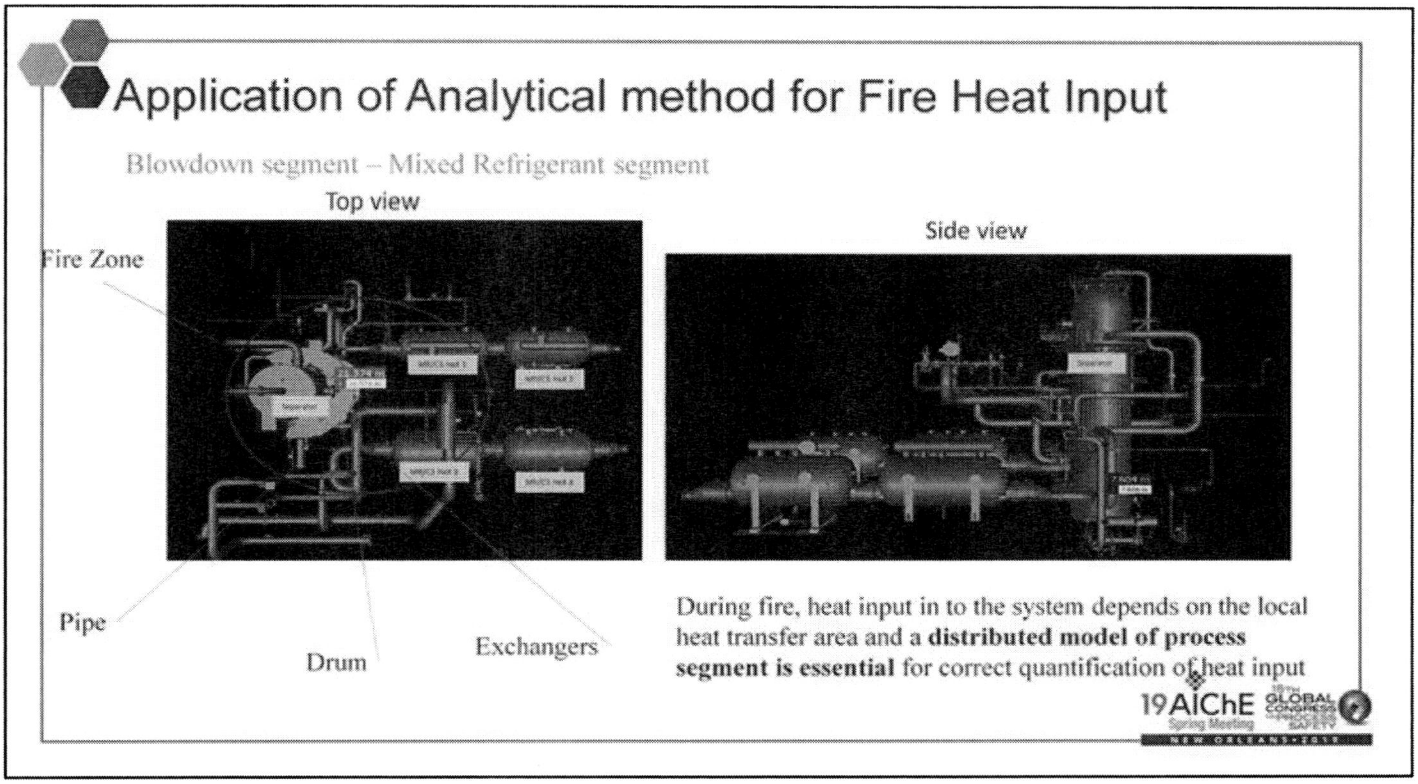

The second thing is that the tool used should be able to take in to account of the distributed nature of blowdown segments. Here is the 3D drawing of the HP Separator blowdown segment, the segment contains the HP Separator drum, 4 exchangers and pipes of different dimensions. As each process element as has its own volume, surface area for heat input, hydrocarbon inventory which could be gas, liquid or both, it is essential that a distributed modelling of the blowdown segment is used for a realistic calculation of the fire heat inputs and depressurisation rate of blowdown segments.

Application of Analytical method for Fire Heat Input

Criteria for Pipe/Drum Rupture

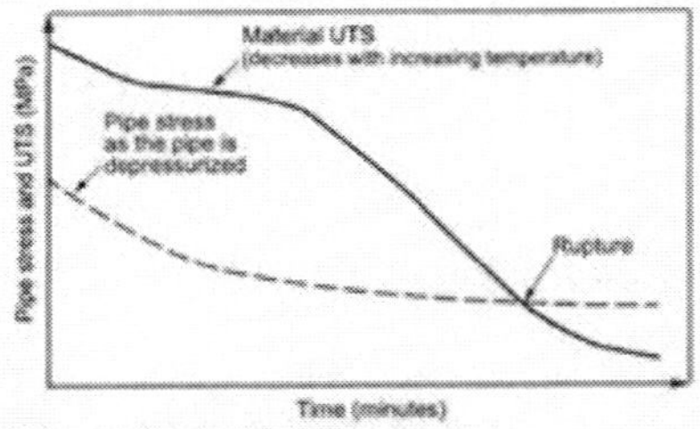

NOTE The pipe will rupture when the pipe stress exceeds the material UTS.

Figure A.3—Illustration of Actual Stress Compared with the UTS for a Fire Exposed Pipe That is Depressurized *API 521, 6th edition*

Rupture is predicted if
Pipe stress > Material Tensile Strength

As I mentioned earlier, application of local peak fluxes provide a conservative calculation of metal wall temperatures and how the material tensile strength is affected by the increase in wall temperature. The global average heat fluxes provide a calculation of pressure rise within the process and in this reflected in the pipe pressure stresses. If the pressure stresses exceeds the material tensile strength, then rupture is predicted

Case Study: Fire Blowdown of Mixed Refrigerant (MR) Separator Segment

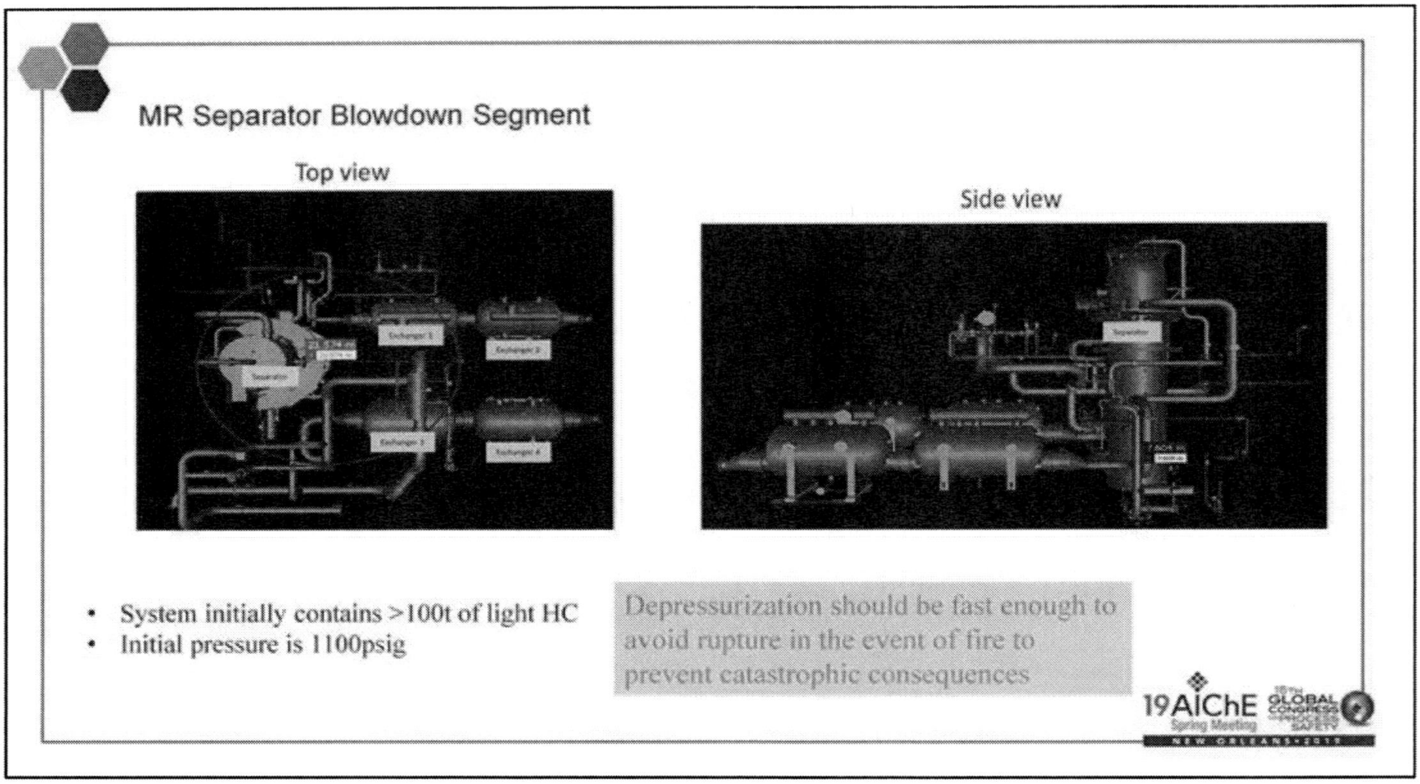

This is a Mixed refrigerant HP Separator blowdown segment in a LNG plant. The MR exchanges heat with C3 refrigerant and is flashed in the HP Separator. The system is at high pressure of around 1100psig and the HC inventory in the system is in the excess of 100tonne. It is important to get the blowdown system correct to ensure that the system depressurise safely in a fire scenario to avoid rupture and fire escalation.
This was a plant that was already constructed and the operator had some concerns about the fire survivability of this blowdown segment

The top view shows the Fire Zone . API 521 specifies that the fire can spread to a maximum of 5000ft2 and this has been taken as the basis in this study. From the Fire Zone is it clear that the segment has a wide plot area and that at any time, only part of the segment would be subject to fire i.e., only the pipe and vessel metal surfaces within in the fire zone.

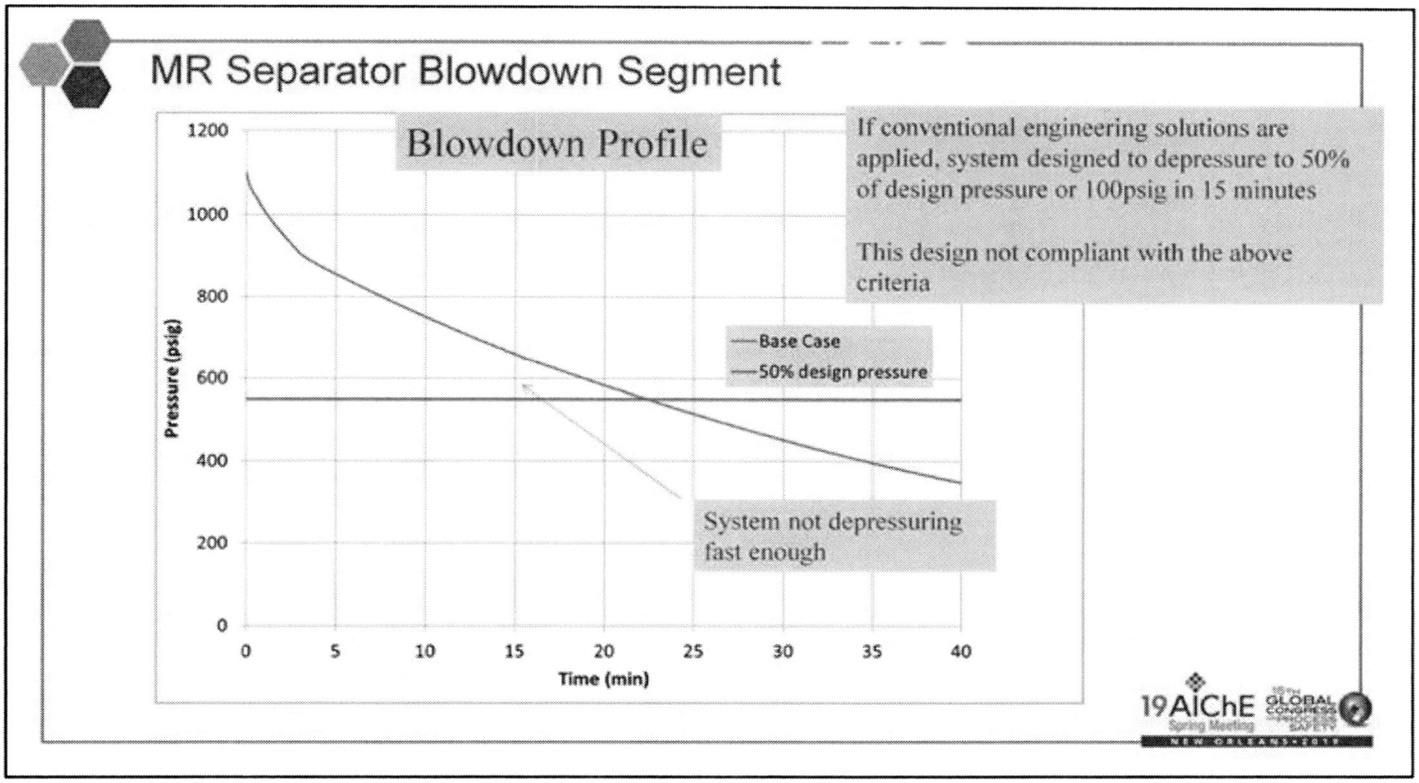

This plot shows the pressure profile of the blowdown segment during the fire scenario. The depressurisation profile is compared against the 50% design pressure.

If the conventional engineering practices are applied, an engineer would design the blowdown segment to depressure in 15 minutes to 50% of design pressure or 100psig, whichever is the lowest. However, in this case, the blowdown calculations showed that the segment would violate the design criteria. At 15 minutes, the segment pressure was higher than the 50% design pressure. To meet the criteria, solution would either be

- to increase the blowdown rate to ensure that the system depressurises faster
- reduce the heat input in to the system by introducing passive fire protection (PFP).

If the plant was in the design phase, it would be easier to take these design changes in to consideration but these are not preferred solutions in the case of plant that has already been built. This is one of the areas where a rigorous fire survivabiliuty calculations provides considerable additional value.

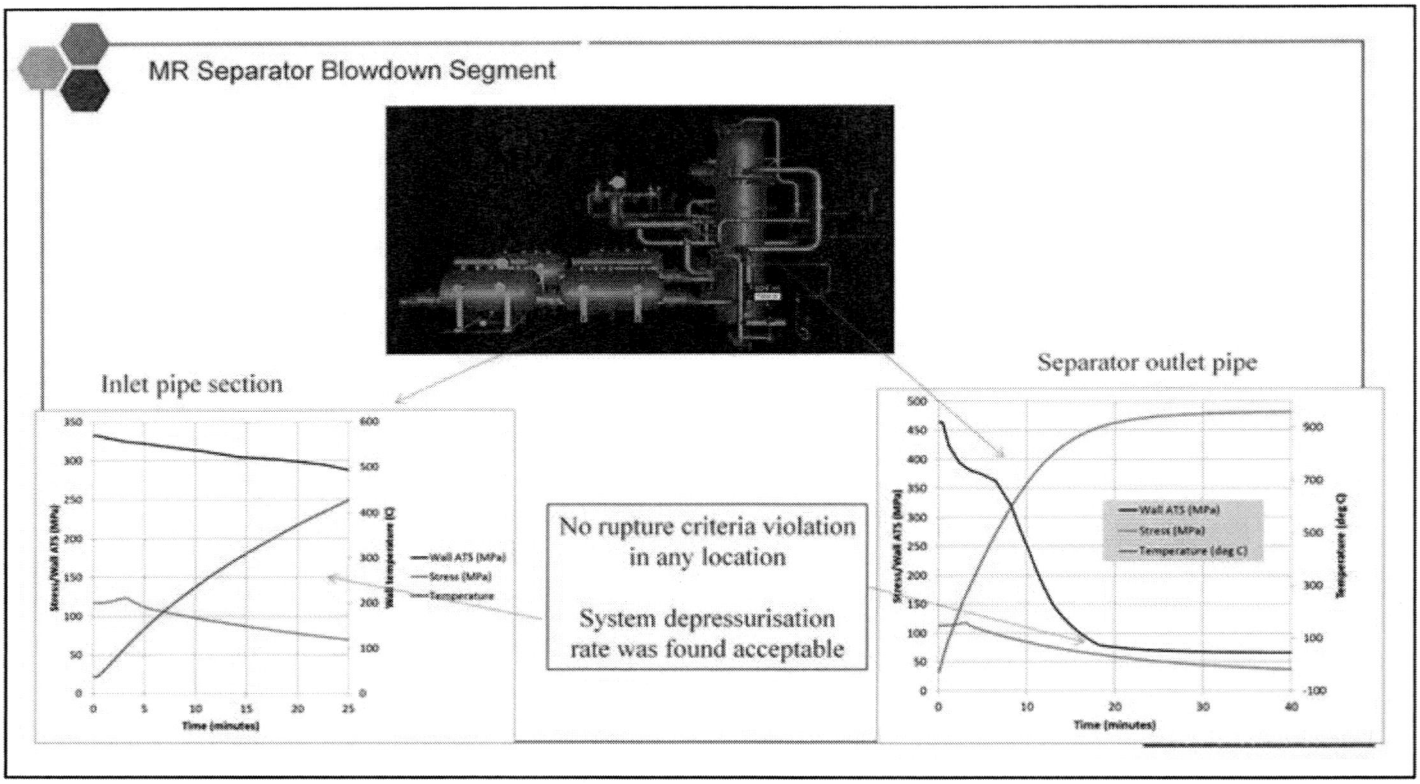

Here are some selected plots from the fire survivability calculations. The plots show the material tensile strength of the metal, pressure stresses and metal wall temperature as a function of time.

If the pressure stresses exceed the material tensile strength then rupture is expected to occur. In this study, we found that in spite of the slower depressurisation, rupture was not expected to occur anywhere within the blowdown segment. So, though the depressurisation rate is slower than the guideline value provided in API 521, it was adequate to avoid rupture within the process segment and design was found to be adequate.

This study helped the operator avoid unnecessary plant design changes which would have been costly and disrupted production.

Summary

- API 521 guidelines for fire blowdown and survivability has been updated in the current version (6th edition)
  - Rigorous analytical methodology suitable for both pool and jet fire assessments
- A distributed blowdown segment modelling approach is required for a realistic analysis of the segment fire survivability
- Fire survivability analysis help avoid unnecessary operating plant blowdown design changes

To summarise, the current version of the API 521 guidelines has been updated for the fire blowdown and survivability assessments. The rigorous analytical methodology for fire heat input is suitable for both pool and jet fire assessments. A distributed modelling approach of blowdown segments is essential for a realistic analysis of the fire survivability of blowdown segments. The case study shown here was carried out using distributed models and the methodology adopted is compliant with the latest API 521 guidelines.

The detailed fire survivability assessments can help avoid unnecessary design changes to operating facilities.

For a comprehensive review of both process segments and flare network, this analysis can be extended to review with dynamic flare loads and check for capacity constraints. Due to time constraints, I haven't discussed the flare side of the analysis but if you are interested, please come and speak to me after this talk.

Design and Operability Analysis of Multi-tank Boil off Gas Compressor System

Subhash Patil and Jaleel Valappil

Bechtel Oil, Gas and Chemicals

Prepared for AIChE Spring Meeting 2019.

Keywords: Dynamic simulation, Boil-off Gas, LNG, Compressor controls

Abstract

This paper presents the design and operability studies of BOG compressor network done on a greenfield facility with multiple tanks and compressors. The results from these simulations were valuable in verifying engineering design, providing operational advice and in enhancing the process control schemes. The various insights gained from these analyses are presented in this paper.

1.0 Introduction

Multi-train LNG facilities are economically attractive option for LNG producers and several of these facilities are in operation all over the world. These facilities use one or more LNG storage tanks and common boil off gas (BOG) compression system that process the BOG generated in the tanks and during ship-loading. In some cases, asymmetrical plant, tank farm and jetty layouts introduce an intrinsic imbalance in operation of the BOG system. In addition, these asymmetrical layouts introduce an operating challenge for a steady and balanced distribution of ship vapor return and BOG generated in LNG tanks. A detailed analysis using dynamic simulation helps in verifying the hydraulic system and control scheme adequacy for various operating scenarios comprising varying number of tanks and number of BOG compressors in operation. The various operating conditions, including tank operating pressures, compressor operating points and other parameters can be obtained from these simulations for engineering design or verification.

Bechtel has successfully used simulation technology for BOG compressor network analysis over the years. For design studies, a fundamental dynamic simulation model is used. This is developed based on the actual equipment sizes, piping isometrics, heat leak calculations and other relevant information. If more rigorous representation of the control scheme is desired, it is possible to utilize an emulator to give an exact replica of the control system.

For the case study presented here, an actual emulation of the compressor control system is connected to the dynamic plant model of the entire BOG system. This is useful to check the anti-surge and master controller performance and identify any potentially undesirable interactions. Various operating scenarios including startup and shutdown of BOG compressors, emergency shutdowns, ship loading, and suction and discharge valve failures were studied.

2.0 Process and Model Description

Process Overview and Significance of Dynamic Simulation Analysis

The case study presented here involves a large LNG facility with multiple LNG trains/plants, LNG storage tanks, loading jetties and BOG compressors. A high-pressure BOG compressor system processes LNG rundown flash vapors generated in these tanks and vapors generated in ships during loading operation. The processed high-pressure BOG is the main source to the fuel gas unit. The excess BOG is recycled to the liquefaction section of the plant.

There was a substantial degree of asymmetry in the layout of LNG refrigeration trains/plants, LNG storage and loading jetties. In addition to the asymmetrical layout, phase wise construction and commissioning of multiple LNG trains needed a detailed analysis of system operation during various phases and scenarios such as startup and trip of BOG compressors, loading of ships, failure of compressor suction valves and other similar scenarios identified during safety reviews.

Simulation Model Description

The BOG simulation model consists of two parts: process model and control model. The two models were integrated using Unisim Operations platform. Unisim Design was used to develop the process model, CCC Series 5 was emulator used to develop the BOG compressor control model and Unisim Operations was used to link the two models.

A dynamic simulation model was developed using standard unit operations available in the software. The model included ship vapor return lines, LNG tanks, BOG compressors, suction drums, discharge air coolers and associated piping. Process equipment details were obtained from the datasheets of respective equipment. Process control information was obtained from vendor provided control and safety system description. Piping and fitting information was taken from the piping isometrics to estimate volumes and resistances of various pipe sections.

The Series 5 emulator from Compressor Controls Corporation (CCC) was used to model the master, load sharing, and anti-surge controllers used in the BOG system. This emulator is an actual representation of the CCC control functionality and it provides exact control responses. The tuning parameters from the BOG simulation study can be utilized as the starting point for the field controller tuning during startup/commissioning.

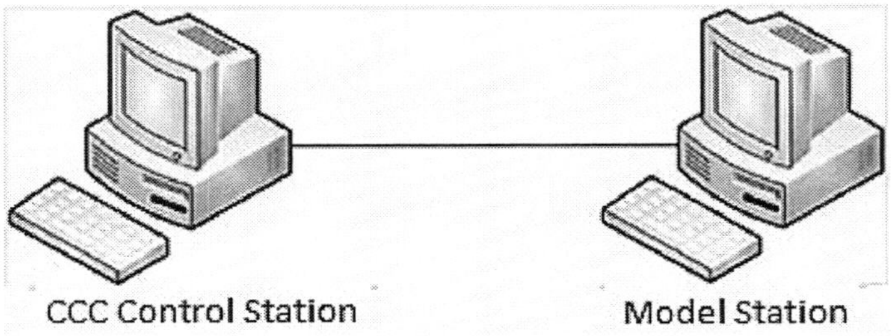

Figure 1: Simulation System Set-up

3.0 Simulation Cases and Results

Start-up of a BOG Compressor for Ship Loading

In this scenario, to prepare for 2nd ship loading, an additional BOG compressor was started. After the started-up BOG compressors were operational and stable, 2nd ship loading was initiated. The stability and controllability of the BOG compression system was analyzed. Tank pressures were monitored, and hydraulic limitations were evaluated. The BOG compressor COMPR-A was started at 15 minutes into the simulation. At 35 minutes, 2nd ship loading was started.

Below is the response of the BOG system upon start of a BOG compressor followed by initiation of second ship loading with the first in progress.

Before implementing recommended changes:

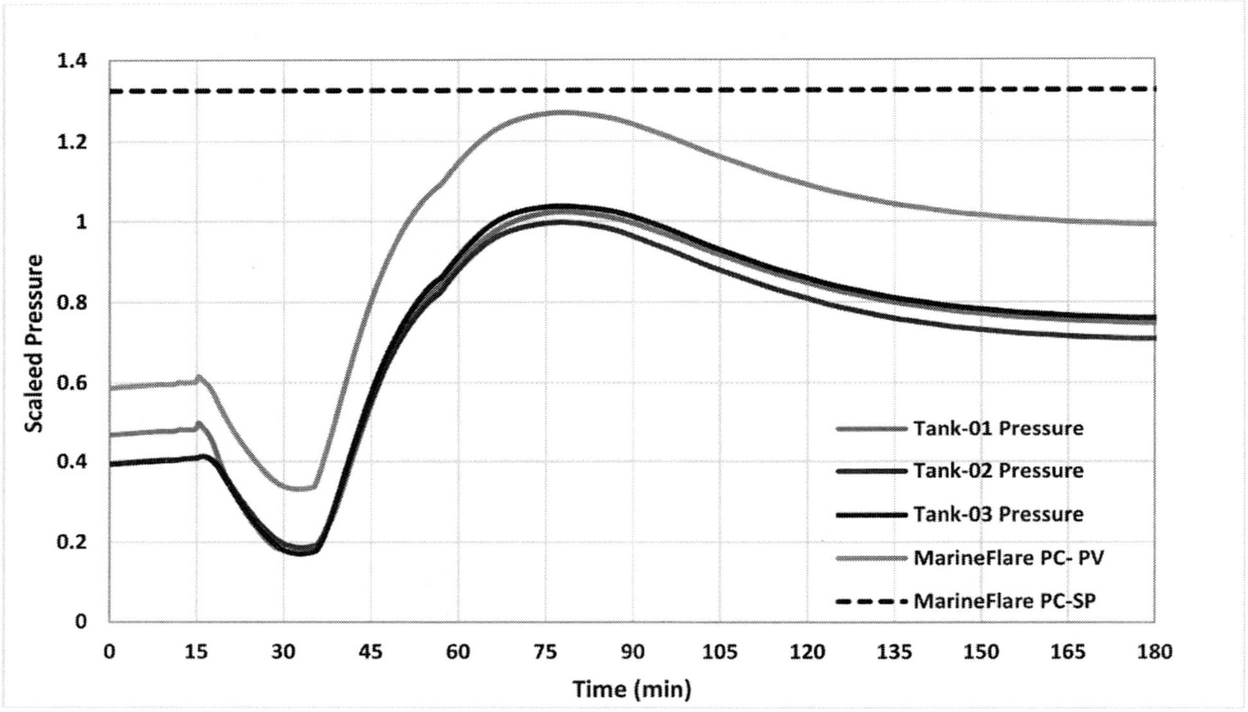

Figure 2: LNG Tank and Ship Vapor Return Header Pressure (Before)

It can be seen from figure 2, upon initiation of 2nd ship loading, the ship vapor return header pressure rose very close to the flare valve set pressure. The tank pressures went close to minimum operating pressure.

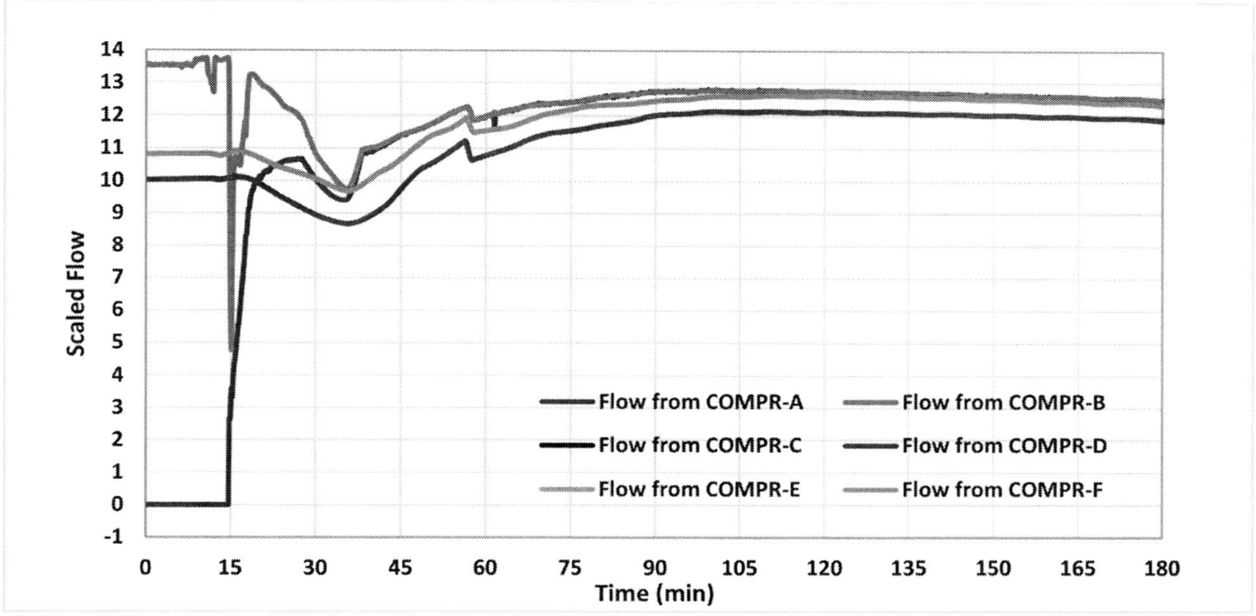

Figure 3: Ship Vapor Return and Marine Flare Flow (Before)

Figure 3 shows that during startup of a BOG compressor, operation of the other compressor in its pair (COMPR-A and COMPR-B here) was affected greatly.

<u>After implementing recommended changes:</u>

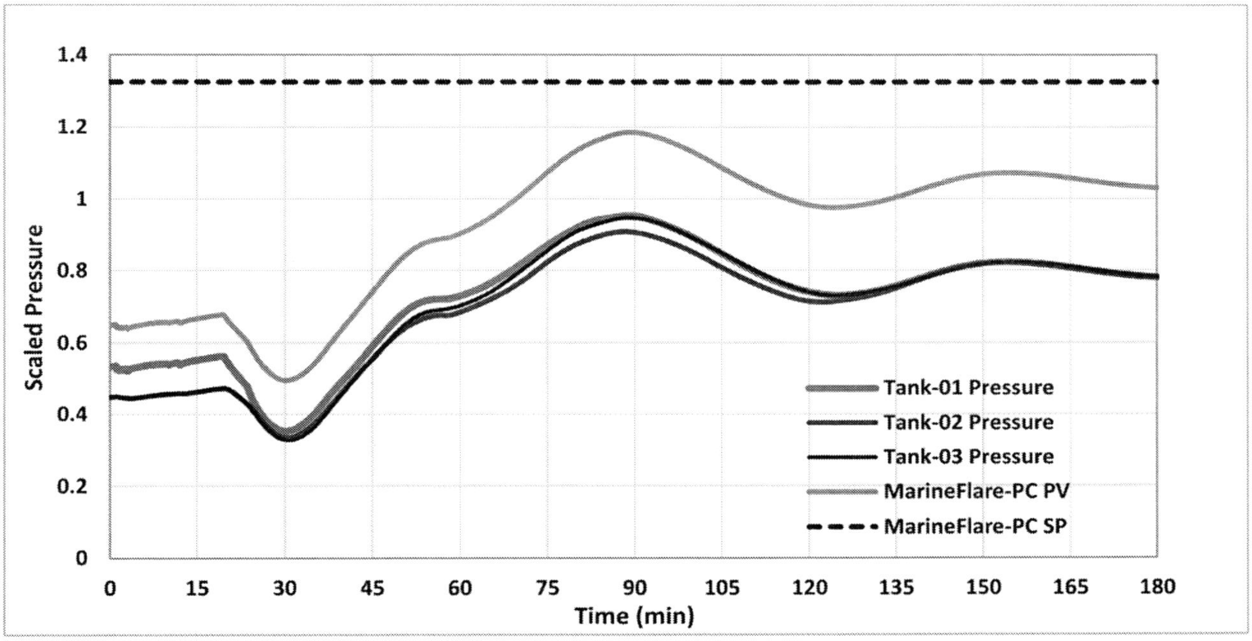

Figure 4: LNG Tank and Ship Vapor Return Header Pressure (After)

Figure 4 shows that after making changes in the control scheme and tuning parameters, the tank pressures were within an intended range.

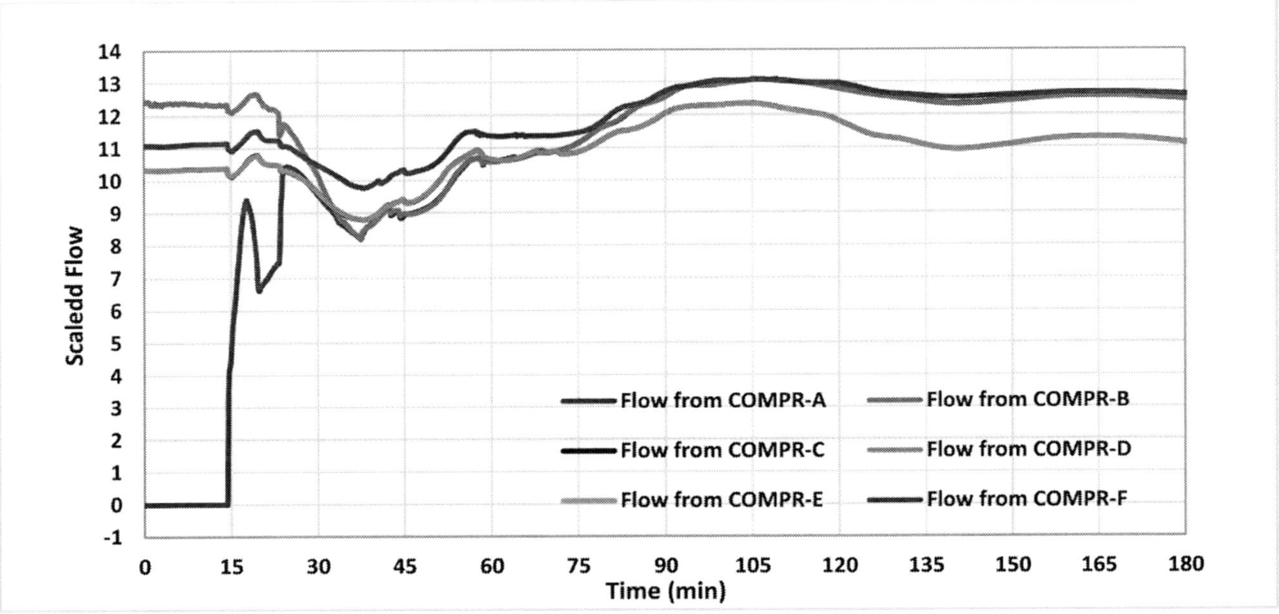

Figure 5: Ship Vapor Return and Marine Flare Flow (Before)

It can be seen from figure 5 that the operation of compressor in a pair was not affected by startup of the other compressor after implementation of recommended changes.

Trip of a Single Operating Compressor

In this scenario, the only operating compressor (COMPR-B in this case) in a pair was tripped. The effect on BOG compression system was observed. The event occurred at 20 minutes into the simulation.

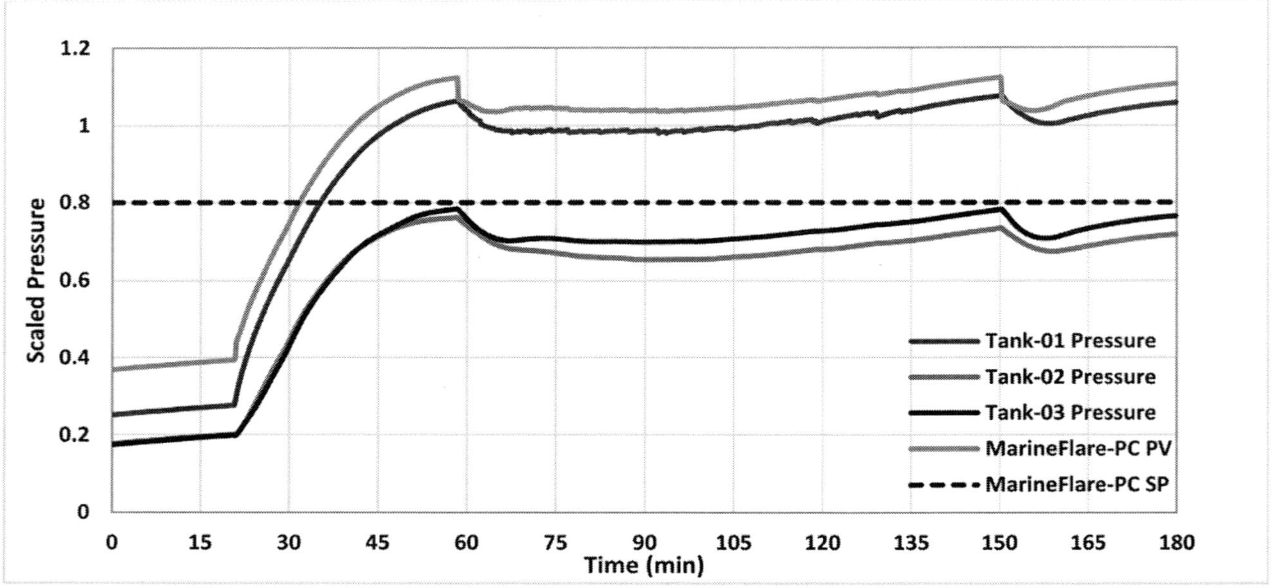

Figure 6: LNG Tank and Ship Vapor Return Header Pressure

Figure 6 shows response of the BOG system upon emergency shutdown of the only operating compressor in a pair. It shows the affected tank pressure was operating very close to the relieving pressure with the flare valve opening intermittently.

4.0 Summary and Conclusions

The study was valuable in verifying the design and operability of the integrated BOG system. It was also useful in providing design and operational guidelines. Some of the recommendations made from the analysis include:

1. Positions of anti-surge and quench valve prior to and during start-up of the compressor.
2. Coupling and de-coupling of recycle balancing for a compressors pair during start-up, shutdown and failure of suction and discharge valves.
3. Location of overpressure control measurement.
4. Optimization of quench usage during BOG compressor recycle.

The benefit from undertaking such analyses at an early stage of the project such as FEED/EPC is to identify the potential operational and design issues and develop effective strategies in advance to avoid start-up delays and operational difficulties.

19 AIChE Spring Meeting
15TH GLOBAL CONGRESS ON PROCESS SAFETY
NEW ORLEANS • 2019
March 31 – April 4, 2019
Hilton New Orleans Riverside

Three Cautions when Designing an MR Compressor Speed Control Scheme

William P. Schmidt
Janet Firley Mitchell
Matthew J. Okasinski, Ph.D.
Jeremy D. Beard
Air Products and Chemicals, Inc.

19 AIChE Spring Meeting
15th Global Congress on Process Safety
New Orleans • 2019

CAUTION

Closed System
Ahead

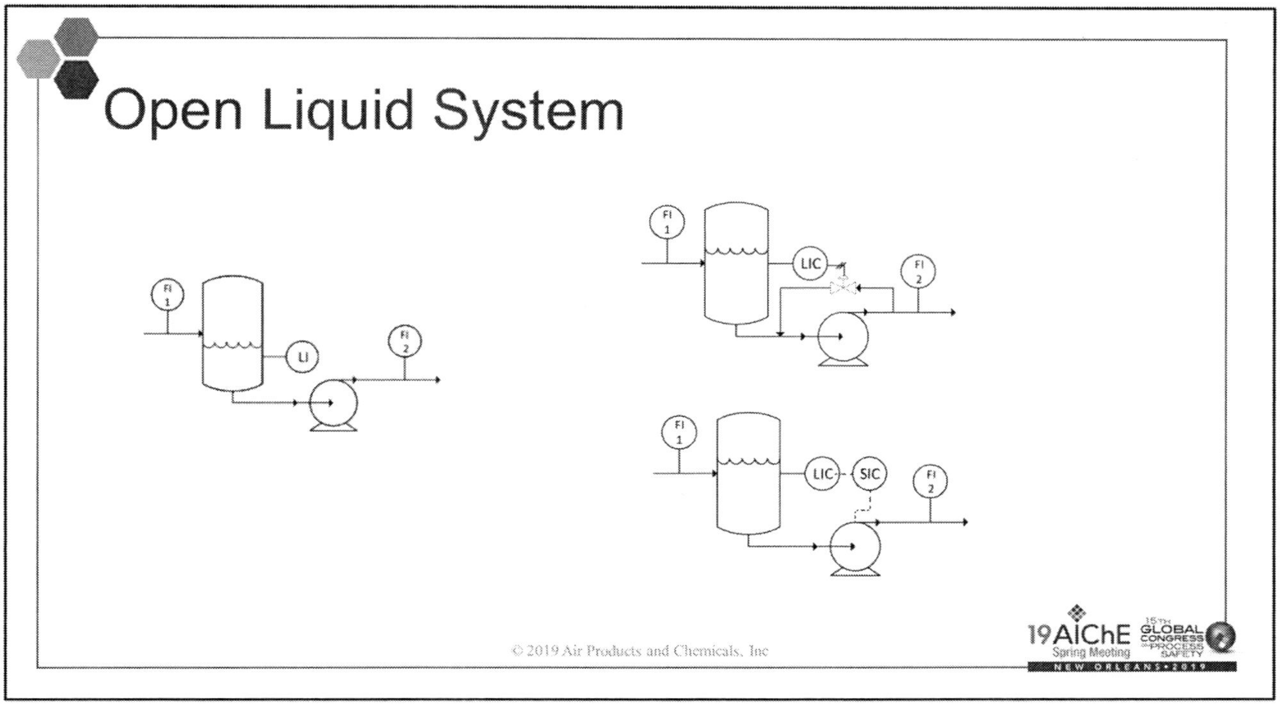

Let's review some simple examples. Liquid flows into a vessel and is withdrawn by a pump. When the inlet and outlet flows are equal, the system is at steady state and the vessel level is constant. What are some ways to control the pump so that the two flows are equal?

One option: The pump recycle valve is manipulated to maintain a constant liquid level. Recycling more liquid will reduce the liquid flow going forward (FI-2) but keep the level constant

Another option: Use speed to control the outlet flowrate; speeding up the pump increases the outlet flowrate; slowing it down decreases the outlet flowrate.

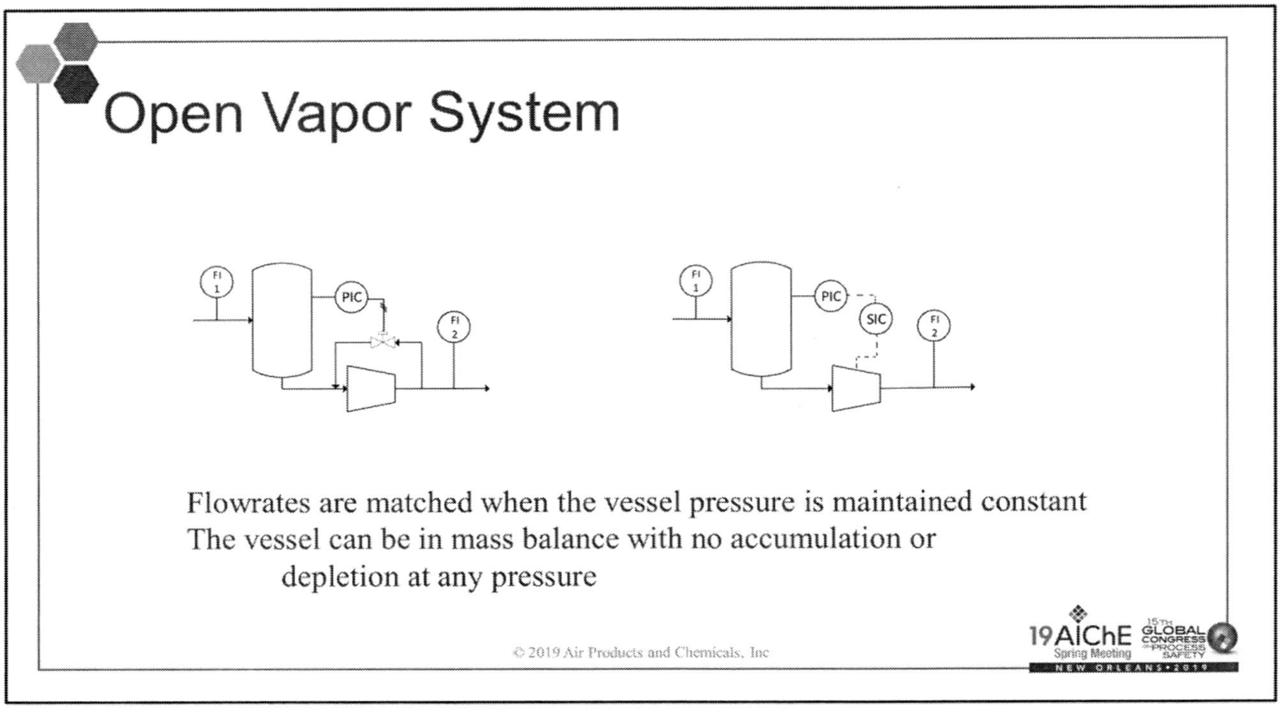

An open vapor system is analogous to an open liquid system. The difference is that inventory is measured by pressure rather than level.

Closed Liquid System with invalid control

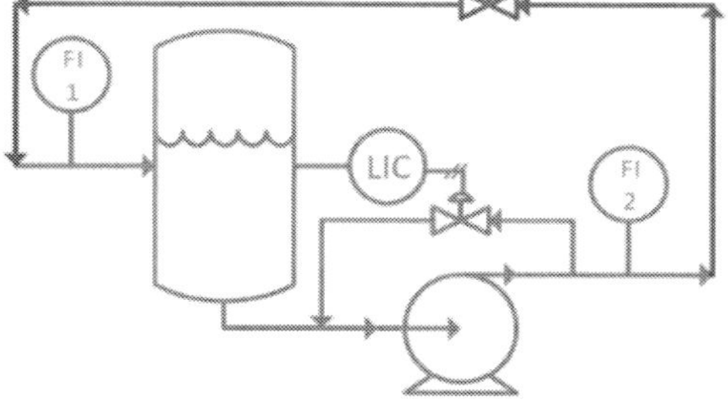

19AIChE Spring Meeting 15TH GLOBAL CONGRESS ON PROCESS SAFETY NEW ORLEANS • 2019

Let's return to our example of an open liquid system but now close the loop with the red line. The pump discharge flow is now always equal to the vessel inlet flow because they are the same stream. This is a closed system with a fixed inventory. Adjusting the pump recycle will change the circulating flowrate but will no longer change the level, which is fixed by the inventory initially charged to the system.

If this fundamental difference between open and closed systems is not recognized, the level controller will move the recycle valve into one of two positions, depending on the choice of the level setpoint.

- If operating level is below the controller setpoint, the controller will attempt to raise the level by opening the pump recycle valve; This will have no effect on the level, so the recycle valve will wind open, resulting in a minimum circulating flowrate
- If the operating level is above the setpoint, the controller will attempt to lower the level by closing the pump recycle valve; Again this will have no effect on the level, so the recycle valve will wind closed, resulting in a maximum circulating flowrate

In other words, a control scheme for an open system may not be suitable for a closed system.

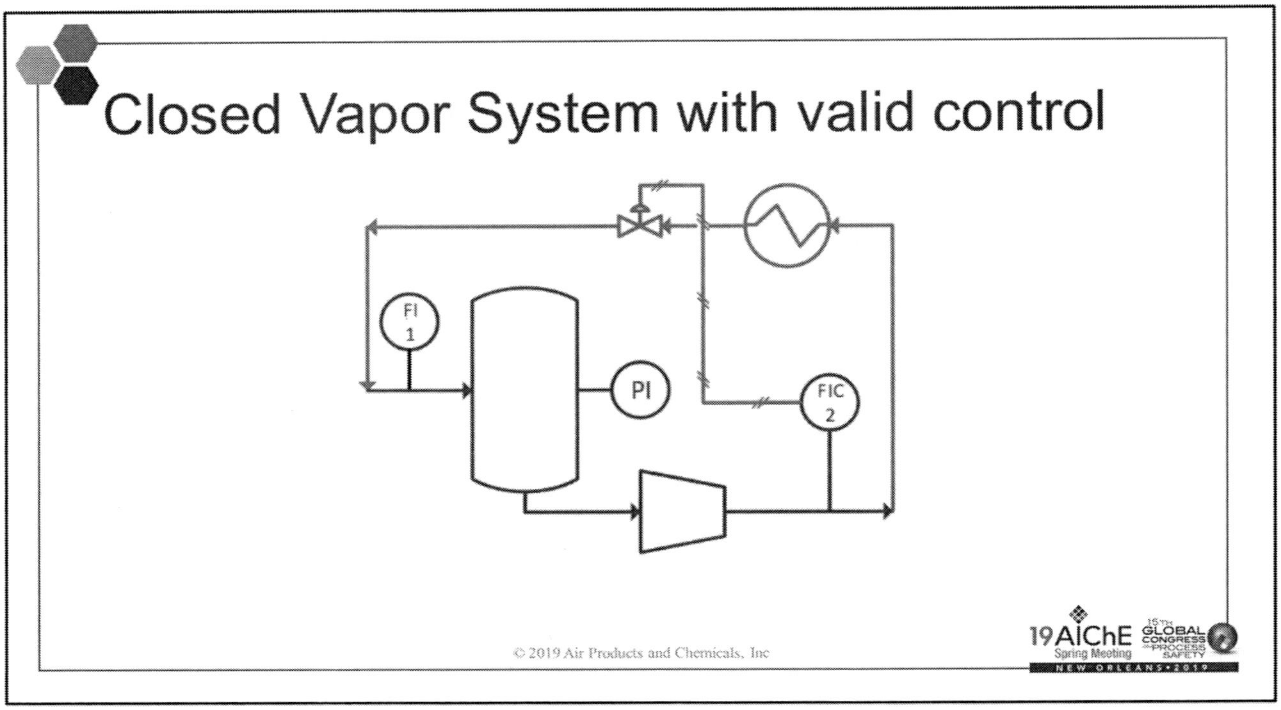

Let's return to our example of an open vapor system but now close the loop, shown in red. A heat exchanger and throttling valve are added remove the heat of compression and reduce the stream pressure back to the suction pressure.

If more flow is needed, FIC-2 will open the throttle valve, reducing the discharge pressure and moving the operating point of the compressor to a lower head, higher volumetric flow. Alternatively, if less flow is required, closing the throttle valve will increase the discharge pressure and move the operating point to higher head, lower volumetric flow condition.

Open and Closed Systems at Steady State

	Open System	Closed System
Inventory	Many	Fixed
Flowrates at Steady State	Fixed (Outlet =Inlet)	Many

These simple examples illustrate a key difference between open and closed systems; they must use different strategies to control flowrates. An open system can have many different inventories but have a fixed flowrate at steady state. A closed system has a single inventory, but many steady state flowrates.

Now that we've discussed these simple examples between open and closed loop control, let's take a closer look at compressor operation.

CAUTION
Watch
Your Head

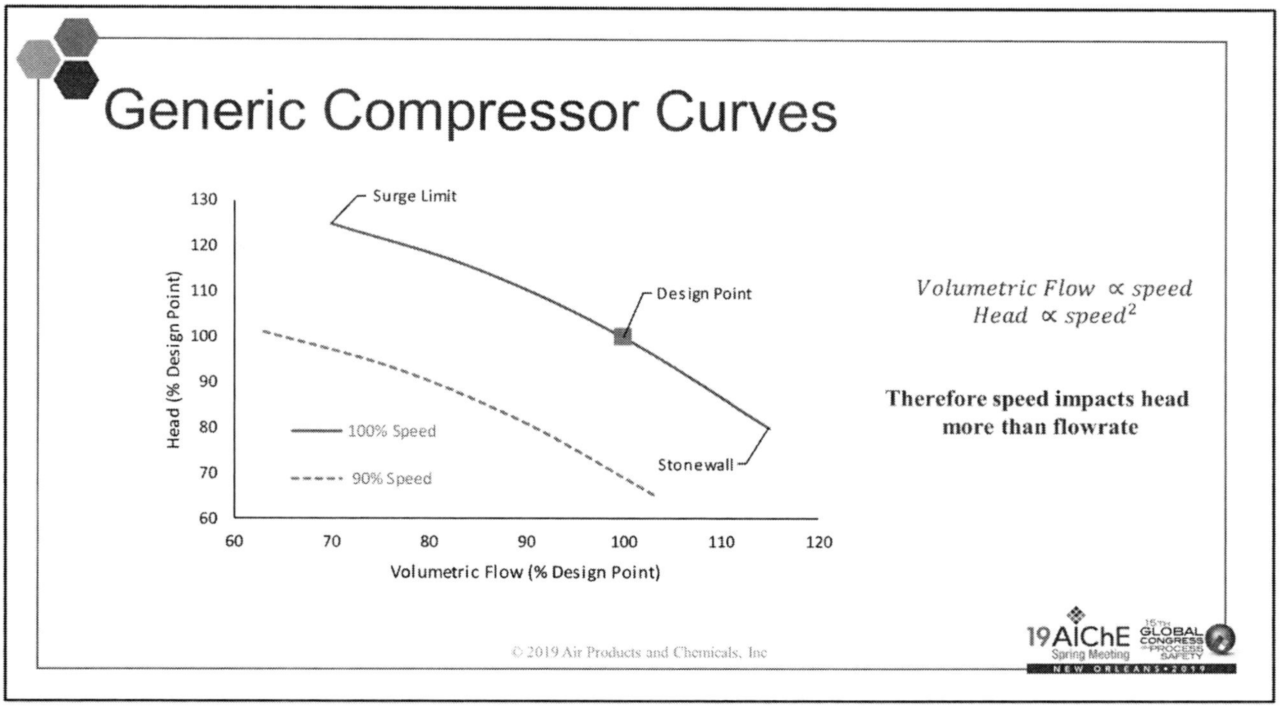

This graph illustrates how speed affects performance. By Fan Laws, the volumetric flowrate increases linearly with speed but head increases with the square of speed. Because the amount of refrigeration supplied depends on the pressure ratio, small changes in speed can greatly affect refrigeration produced. Therefore in the closed MR system, it is more accurate to think of speed primarily affecting head rather than flowrate.

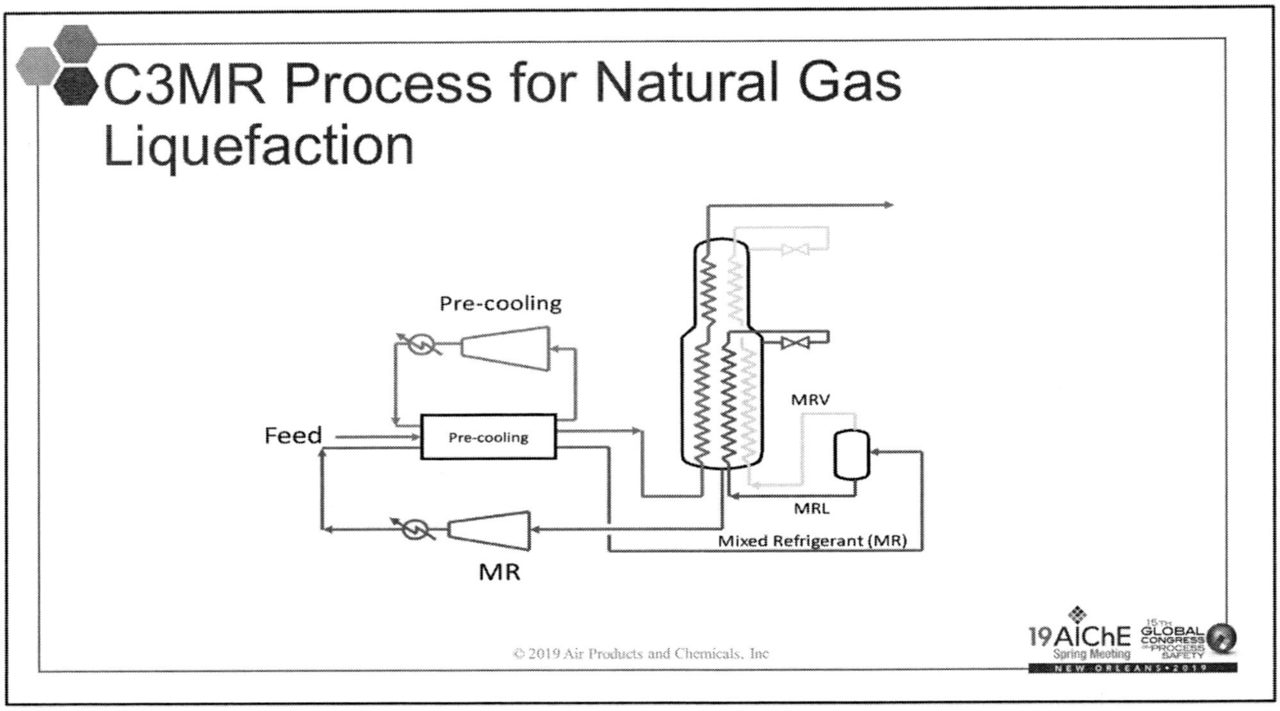

To review, the clean dry feed is first precooled, typically by propane, then further cooled, liquefied and subcooled against a mixed refrigerant in the Main Cryogenic Head Exchanger (MCHE).

The propane is compressed, cooled and condensed before being let down in to several pressure levels to provide cooling duty for the feed and mixed refrigerant streams.

The mixed refrigerant is compressed, cooled and then flashed into a liquid (MRL) and vapor (MRV) streams. These are further cooled and liquefied in the MCHE. The MRL is taken off first, let down in pressure and then used aa refrigerant on the shellside of the lower portion of the MCHE. The MRV is taken off at the coldest point in the MCHE, let down in pressure across a Joule Thomson valve and used as refrigerant on shellside throughout the MCHE.

The liquefaction process consists of both open and closed loop systems. The feed stream is an open system; whereas the propane and mixed refrigerant circuits are closed systems with compressors.

CAUTION

Avoid Suction Pressure Control
for MR Compressors

Defining the MR System

Input	Set by
Inventory	Operator
Circulating MR flowrate	Determined by LNG production; set with JT valves
Compressor Speed	Industrial Gas Turbine – fixed Aeroderivative Gas Turbine – possible to vary

As we saw earlier in the open vapor system, the compressor suction pressure can be held constant by adjusting the compressor flowrate. However in the closed MR system, the MR compressor suction pressure is a function of many variables such as:

Total MR inventory, MR composition, compressor speed, the overall LNG process operation (actual production rate and LNG temperature) and of course compressor flow. Thus it is difficult to control suction pressure based on a single parameter.

The closed MR system can be defined by 3 inputs: Once these are chosen, all other operating parameters are fixed.

Dynamic simulation results

Case		DS-1	DS-2
Inputs	LNG Production	100%	85%
	Speed	100%	100%
	Inventory	Same	Same
Results	Suction Pressure (bara)	4.28	3.79
	Discharge Pressure (bara)	44.8	45.7
	Compression Ratio	10.5	12.1
	Power to recycle	0.0%	4.5%
	MR Spec Power (relative)	100%	106.8%

19 AIChE Spring Meeting 15th Global Congress on Process Safety NEW ORLEANS • 2019

Dynamic simulations were performed to rigorously model a constant inventory system. The key inputs to the simulation, as described on the previous slide, were LNG production, speed and inventory. The first two cases run were the base case (DS-1) and a turndown to 85% production at constant speed (DS-2).

Notice that the suction pressure decreased and the discharge pressure increased, causing the compression ratio to rise about 15%. At this lower production level, the MR compressors must recycle to prevent entering the surge region so there is power used that doesn't produce LNG, thus the specific power rises about 7%.

Dynamic simulation results

Case		DS-1	DS-2	DS-3
Inputs	LNG Production	100%	85%	85%
	Speed	100%	100%	95%
	Inventory	Same	Same	Same
Results	Suction Pressure (bara)	4.28	3.79	4.45
	Discharge Pressure (bara)	44.8	45.7	43.9
	Compression Ratio	10.5	12.1	9.88
	Power to recycle	0.0%	4.5%	2.7%
	MR Spec Power (relative)	100%	106.8%	103.8%

Now let's examine lowering the speed at turndown.

While the compressors still need to recycle, they are on a lower speed curve than DS-2, producing less head so the compression ratio is lower as is the power spent on recycling. Thus the specific power, while still higher than the base case, is lower than using design speed at turndown production.

Note that comparing the two turndown cases (DS-2 and DS-3), the speed decreased only 1/3 as much as production. 15% production change to only 5% speed change. As mentioned before, speed primarily influences head rather than flowrate.

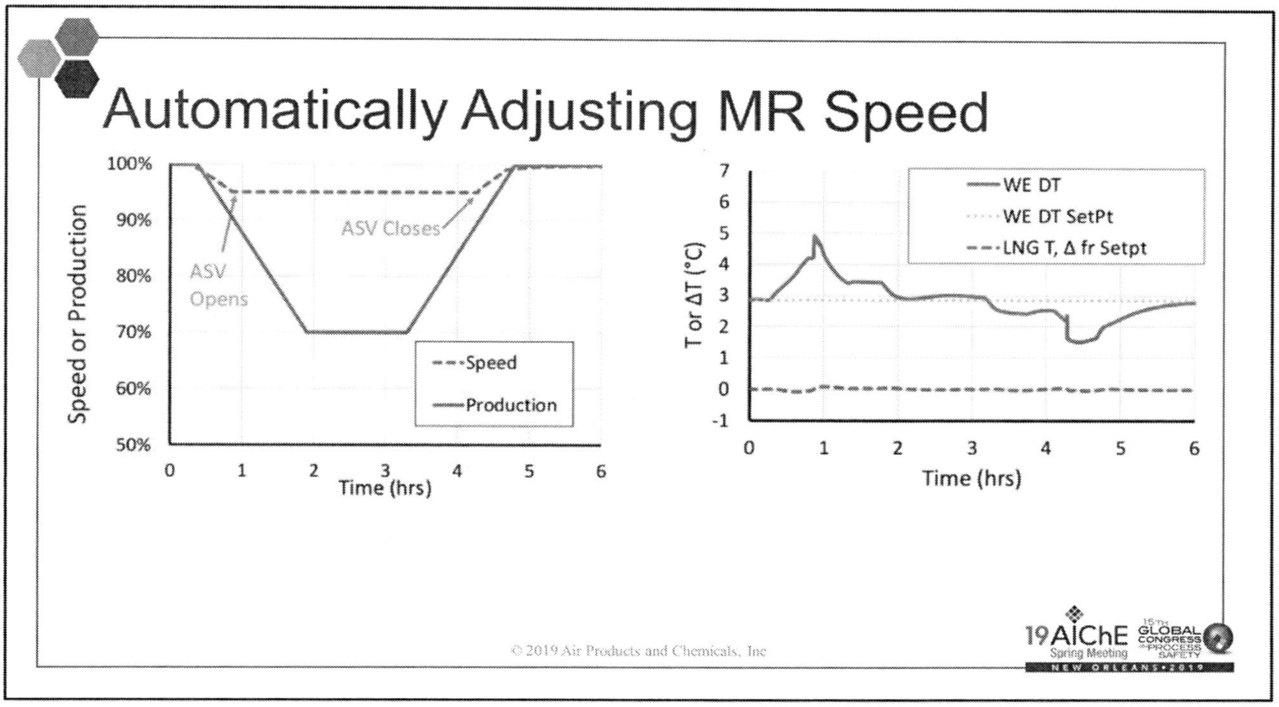

A dynamic simulation was run with the Air Products Enhanced Control Scheme automatically ramping the plant production from 100% to 70% and back to 100%. The Enhanced Control scheme was modified to ratio the speed with the refrigeration flowrate. When the anti-surge valves open, the speed setpoint is held constant to prevent unwanted interactions between the surge protection system and speed control.

With the MR inventory fixed, the control scheme automatically adjusted the JT valves and speed to keep the key operating parameters within the desired range: namely production, LNG temperature and MCHE warm end temperature difference which is a measure of efficiency.

The dynamic simulations confirmed that the modified control scheme could adjust speed while varying the production, keeping the LNG temperature on setpoint and operating the MCHE efficiently.

Conclusions

- Three facts to be considered to effectively control the MR system:
  - MR system is closed; open system control does not apply
  - Varying speed affects refrigeration much more than flow
  - MR suction pressure cannot be controlled by speed alone
- In LNG liquefaction
  - Gas turbine – don't vary speed due to inherently small speed range
  - Aeroderivatives – vary speed based on required refrigeration

550658 Multi-Variable LNG Optimization Using Parallelized Simulator with Integrated Optimizer

LNG liquefaction is an energy-intensive process, and thus lends itself to optimization. Our work highlights the propane/mixed refrigerant (C3MR) section of this process. A common practice for C3MR optimization is to connect a generic optimization engine to a process simulator. While this approach can work well for adjusting a single type of independent variable, attempting to adjust many variables can lead to local optima, trapping the optimization engine.
Using CHEMCAD's integrated optimization engine and multi-core technology, we can easily set up rigorous optimizations, run them efficiently, and have a high degree of confidence that our solution is the global optimum.

550574 Controlling Deactivation of ZSM-5 Catalysts in the Methanol-to-Hydrocarbons Reaction By Co-Feeding Water

The methanol-to-hydrocarbons (MTH) reaction yields a wide range of hydrocarbons of great interest in the petrochemical industry such as ethylene, propylene, aromatics and a gasoline-like mixture. This reaction takes place in the presence of a shape-selective acid catalyst, such as ZSM-5 or SAPO-34 zeolites. This kind of catalysts presents a rapid deactivation due to the formation and deposition of coke that blocks the access of reactants to the catalyst active sites. However, the catalyst can recover its activity after being subjected to a regeneration process by burning the coke deposited on it. Thus, the MTH catalyst may operate in reaction-regeneration cycles. Many efforts have been made with the purpose of improving the catalytic performance of the MTH reaction with special emphasis on attenuating catalyst deactivation. One common strategy is co-feeding water, which has been proved to be satisfactory for increasing the lifetime of SAPO-34 catalysts with an important impact for the industrial implementation of the methanol-to-olefins process.[1] However, this effect has not been studied with such detail for ZSM-5 catalysts, whose major research efforts are focused on zeolite modifications. Indeed, the modifiers improve the overall catalytic performance of ZSM-5 catalysts, such as delaying the catalyst deactivation that allows longer operational periods.[2,3] In an earlier stage of our research, we observed that the modification of ZSM-5 zeolites with $ZnCl_2$ improves the catalytic performance of the MTH reaction by reducing the formation of coke that leads to delay the catalyst deactivation. Now we aim to further improve the ZSM-5 catalyst performance in the MTH reaction by co-feeding water based on the experience reported for its counterpart SAPO-34.

In this work, we investigated the effect of co-feeding water with methanol as a strategy to control the deactivation of a Zn-modified ZSM-5 catalyst in the MTH reaction. The MTH experiments were performed in a fixed-bed reactor at 400 ºC, using different water contents in the feed (0, 10, 20 and 33 wt%), 100 mg of catalyst and a methanol flowrate of 0.1 (mol of carbon) h^{-1}. Each MTH experiment was run until the catalyst was highly deactivated (conversions under 5%) and the run time was reported as the catalyst operational time. After each run, the catalyst was taken out and subjected to thermogravimetric analysis (TGA) and soluble coke extraction to gain information about the coke retained. Additional MTH experiments were conducted in UV-VIS-NIR and FTIR spectroscopic operando reactors in order to gain more information on the coke formation kinetics.

Our results show that co-feeding water decreases the initial conversion of the methanol and dimethyl ether in this order: 98% conversion (0 wt% H_2O) $<$ 80% (10 wt%) $<$ 29% (20 wt%) $<$ 11% (33 wt%). At the same time, the addition of any water content in the feed prolongs the catalyst operational time from 19 to 26 h on stream (see Figure 1). We further analyzed the catalyst deactivation in terms of the evolution of the conversion against time on stream for each MTH run. This analysis reveals that the conversion drop is slower as the water content in the feed increases. Also, the product distribution analysis in terms of the yield of different products against conversion shows no significant differences regardless of the water content used in the feed. We only observed that the yield of secondary products such as methane and carbon dioxide decreases with increasing water contents in the feed. The content of coke in the spent catalysts decreases as the content of water in the feed increases, following this order: 47 mg g^{-1} (0 wt%) $>$ 26 mg g^{-1} (10 wt%) $>$ 17 mg g^{-1} (20 wt%) $>$ 11 mg g^{-1} (33 wt%). This is congruent with the decrease of the conversion capacity and the catalyst deactivation rate with increasing water contents in the feed.

Figure 1. Conversion evolution against time on stream for different water contents in the feed in the MTH reaction at 400 ºC.

We further studied the formation of retained species in the catalyst by means of UV-VIS-NIR and FTIR spectroscopies in operando reactors under different water contents in the feed. Both techniques allowed us to identify retained species that are reaction intermediates and coke precursors. The results show insignificant differences in the nature of retained species regardless of co-feeding water but significant differences in their concentration and kinetics: lower amounts and slower kinetics with increasing water content. For example, we observed the raise of a UV-VIS band at 394 nm attributable to polyalkylated aromatics[4] for three different water contents in the feed with the following maximum absorbance at 750 s on stream: 0.28 (0 wt%) > 0.24 (33 wt%) > 0.22 (50 wt%). We observed a similar behavior between polyakylated aromatics and polyaromatics in the operando reactors, which shows that the formation of intermediates and coke is attenuated by co-feeding water. These results concord quite well with the observations obtained in the fixed-bed reactor. Thus, the lower conversion is linked with a slower catalyst deactivation and slower coke formation with increasing water contents in the feed.

The use of FTIR spectroscopy also allowed us to correlate the polyakylated aromatic and polyaromatic species with the level of accessibility of acid sites. In fact, hydroxyls groups interacting with Si (silanol), are the most affected ones among the different acid sites. This observation suggests that the extra water co-fed strongly interacts with these sites and this hinders the adsorption and further reaction of reactive hydrocarbon species. Thus, our results prove that the competitive adsorption of water lowers the catalyst surface coverage by any other hydrocarbon species and this decreases the rate of any kinetic event related to the MTH reaction.

In summary, we demonstrated that co-feeding water attenuates the deactivation of the Zn-modified ZSM-5 catalyst in the MTH reaction and this is mainly due to the competitive adsorption of the extra water co-fed. This decreases the amount and formation rate of coke. We also verified that co-feeding water has insignificant effects on the MTH chemistry, as the product distribution against conversion and the nature of intermediates are quite similar regardless of the water content in the feed. From a process engineering perspective, the addition of a small amount of water (10 wt%) in the feed may represent a positive outcome to the development of the MTH process using a Zn-modified catalyst. Under this scenario, the reactor may operate for a longer period reducing the reaction-regeneration cycles and keeping a high production level of primary products.

References

1. P. Tian, Y. Wei, M. Ye and Z. Liu, ACS Catal., 2015, 5, 1922–1938.
2. H. E. H. E. van der Bij and B. M. B. M. Weckhuysen, Chem. Soc. Rev., 2015, 44, 7406–7428.
3. X. Niu, J. Gao, Q. Miao, M. Dong, G. Wang, W. Fan, Z. Qin and J. Wang, Microporous Mesoporous Mater., 2014, 197, 252–261.
4. K. De Wispelaere, C. S. Wondergem, B. Ensing, K. Hemelsoet, E. J. Meijer, B. M. Weckhuysen, V. Van Speybroeck and J. Ruiz-Martínez, ACS Catal., 2016, 6, 1991–2002.

551174 Demonstrating Sandwich Packing Performance in Butadiene Extraction Units

The application of Sandwich Packing, a column packing technology developed by BASF, offers an efficient alternative when increasing the hydraulic capacity of a plant without detrimental effects in terms of separation efficiency. Sandwich Packing is particularly of interest for the capacity increase of existing columns.

This work addresses the successful performance of the Sandwich Packing Technology in a Butadiene Extraction Unit of full-commercial scale. Experience and results from the operation of several years at BASF's Butadiene Extraction Plant in Germany is now available and will be presented.

551331 Kinetic Modeling and Reaction Pathways of C2 to C7 Olefin Transformation over an H-ZSM-5 Catalyst

The interconversion reactions of olefins on zeolite catalysts are important in many petrochemical processes such as, e.g., catalytic naphtha cracking or in methanol downstream processes for the production of light olefins such as the MTO or MTP processes [1, 2]. Due to the complexity of the reaction mechanism and a high number of involved components it still remains challenging to establish reliable kinetic models suitable for process design, particularly in terms of applicability for reactor optimization purposes.

This work aims at kinetic modelling based on extensive own experimental measurement data. We propose a novel reaction network scheme for the interconversion reactions of the olefins including cracking, oligomerization and disproportionation reactions. Additionally, a reaction kinetic model for the formation of side products has been developed based on a new mechanism to precisely describe the hydrogen transfer reactions.

Reaction kinetics of C2 to C7 olefin cracking over a commercial H-ZSM-5 catalyst were studied in detail using a Berty-type reactor. The reactor concept allows for gradient free kinetic measurements and therefore is an excellent choice to establish intrinsic reaction kinetic models.

To investigate the behavior of olefins, dosing of pure (individual) olefins as well as different olefin mixtures was performed over a wide range of conversion. The structure of the reaction mechanism of olefin cracking was systematically analyzed to identify the best suitable model, taking into account both, the reaction network and the reaction kinetic model approach. The numerical regression of the kinetic parameters was carried out using state-of-the-art solvers. The formation of side products was analyzed in detail using a custom-made gas chromatograph which was capable of separating and analyzing more than 100 hydrocarbon species. Based on this high degree of information for the product distribution, a new reaction mechanism for modeling the hydrogen transfer reactions has been proposed in this work. Considered side products include paraffins, naphthenes, cycloolefins and BTX aromatics.

1. Liu, D., et al., *Inter-conversion of light olefins on ZSM-5 in catalytic naphtha cracking condition.* Catalysis Today, 2014. **226**: p. 52-66.
2. Huang, X., D. Aihemaitijiang, and W.-D. Xiao, *Reaction pathway and kinetics of C3–C7 olefin transformation over high-silicon HZSM-5 zeolite at 400–490°C.* Chemical Engineering Journal, 2015. **280**: p. 222-232.

550547 Novel Energy-Efficient Configurations of LNG Cascade Refrigeration Cycles Based on Structural Modifications

Complex and energy-intensive cascade refrigeration cycles are employed for the production of liquefied natural gas (LNG). The cost of the shaft work energy required for refrigerant compression usually dominates the overall operating costs of the LNG plant. Three novel cascade cycle configurations are developed in this work to be competitive – in terms of energy-efficiency – against relevant commercial processes, i.e. the Propane Precooled Mixed Refrigerant (C_3MR) cycle, the Dual Mixed Refrigerant (DMR) cycle, and the Phillips Optimized Cascade cycle. The novel cascade cycle configurations are developed from structural modifications based on the CryoMan SMR process.

To evaluate the competitiveness of the novel configurations, the operating variables of both the commercial and the novel cascade configurations are optimised to minimise the specific shaft work energy demand for refrigerant compression (i.e. total shaft work demand per unit production of LNG, $MW{\cdot}MTPA^{-1}$). The optimisation strategy consists in applying a stochastic algorithm (using Genetic Algorithm) to search the solution space, and the best solution obtained is then used as the initial guess for a deterministic optimisation using SQP. Moreover, the complexity of all the cascade configurations is kept to a similar level (by fixing the total number of compression stages, major contributors to capital cost) to have a 'fair' comparison of their corresponding energy performance.

The three novel CryoMan options are demonstrated to be more energy-efficient than the commercial cascade cycles in a case study for large scale production of LNG. The novel configurations achieved up to 15.5% savings in shaft work energy demand compared to the commercially dominant cascade cycle (i.e. the C_3MR cycle). Compared to the most energy-efficient commercial cascade cycle (i.e. the DMR cycle), the performance of the novel configurations is equivalent to around 6.9% savings in shaft work energy demand. Furthermore, the energy savings achieved by the novel configurations come at the expense of only relatively minor structural modifications compared to the configuration of the DMR cycle.

550766 A New, Powerful Technology: Integrated LNG to Power

The development of a diversified competitive global LNG market has given rise to small to mid-scale applications of LNG. LNG to Power projects have become particularly attractive for developing markets, where traditional infrastructure is underdeveloped. In this context an integrated power solution that is simple, efficient, and cost effective has a distinct advantage over more complex traditional designs.

This paper will present a novel technology for the regasification of LNG by utilizing low-grade waste-heat. The new technology is well suited for integration with power production offering a simple and compact design that is efficient and environmentally friendly. We will compare this technology to the typical options for regasification in LNG to Power projects.

The technology is inherently scalable allowing application for smaller scale installations that are increasingly popular as LNG becomes a cost effective commoditized fuel that is preferred for its low emissions. The ability to recover low grade heat from the combustion of the LNG, increases the efficiency of the system. Because the heat is recovered from the process, the technology is insensitive to environmental conditions, a weakness of alternative methods such as an ambient air vaporizer that is typically used in small scale applications.

This novel technology differs from common intermediate fluid vaporization methods because it eliminates the pumping circuit typically used. The elimination of the pumps reduces maintenance and operating expense. Simplification of the operating and maintenance requirement is important because many of the target applications are in locations where limited infrastructure drastically increases the cost of maintenance.

The applications for this regasification technology is not limited to LNG to Power projects, but can also be used in other services where a sufficient supply of low grade waste heat is available. This includes industrial applications such as steel mills, drying furnaces and baking furnaces. These are unique applications that demonstrate the flexibility of the waste heat based technology.

The new technology presented in this paper for vaporizing LNG using a low grade heat source is perfectly matched to the needs of LNG to Power projects. The stack gas from the power plant provides an ample supply of waste heat. Utilizing this waste heat improves the overall efficiency and lowers the environmental impact of the power generation installation.

MAINTAINING OPTIMUM TRAINS OPERATION MODE TO ACHIEVE SPECIFIED LNG PRODUCT QUALITY DURING LEANER FEED GAS ENTRANCE IN BADAK LNG

Fajar Singgih Kurnia Putra

Process Engineer, Technical Department, Process Engineering and Energy, Badak LNG, Bontang, Indonesia

Corresponding Author's Email: fajars@badaklng.co.id

Introduction

Badak LNG Plant, designed to process rich feed gas, is currently facing operational challenge due to leaner feed entrance. This condition is causing a decrease of the LNG heating value (HHV) below the LNG sales contract specification. As a countermeasure, some amount of LPG were reinjected to the feed gas stream to maintain the LNG product quality as specified.

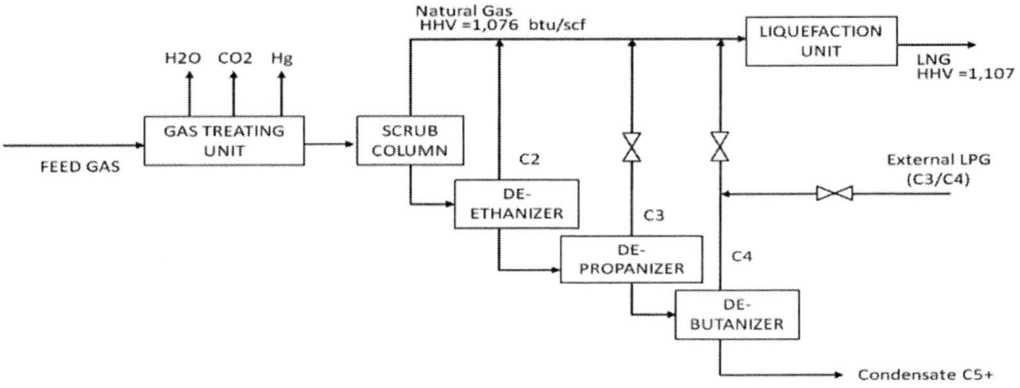

Figure 1. HHV Control Scheme

In order to optimize LPG consumption for this purpose, some improvements have been implemented. One the effort is separating process trains operating mode to produce two LNG qualities, known as dual HHV mode. Rich LNG (HHV $\geq$ 1,107 btu/scf) is produced by Rich Trains, while lean LNG (HHV $<$ 1,107 btu/scf) is produced by Lean Trains.

Implementation of dual HHV mode has been optimized by increasing feed gas temperature inlet Scrub Column of Rich Train to enhance C2+ content at top product, setting the LNG product temperature warmer in order to increase LNG's HHV, and transferring propane product from Lean Train into Rich Train through existing propane interconnecting line.

Meanwhile, Lean Trains have been optimized to enhance C_3 and C_4 extraction by decreasing feed gas temperature inlet Scrub Column in order to extract more C2+ from bottom product of Scrub Column and increasing temperature of Debutanizer to increase C_4 production. Thus, more C_3 and C_4 product can be transferred to increase LNG's HHV of Rich Train.

Objective

The study is intended to evaluate optimum Trains operation mode by implementing several operational optimizations during Leaner Feed Gas Entrance in Badak LNG in order to minimize LPG reinjection consumption.

Evaluation

Dual HHV Operational Mode

Before leaner feed gas entrance, Badak LNG produced LNG product with similar quality/specification in 4 Process Trains. The process scheme was applied in order to maintain LNG quality, mainly heating value and density, within the allowable contract specification.

Actually, based on LNG contracts beyond 2018, mainly there are two groups of LNG specification. The specification can be divided into two major grades, specifically for the heating value and density parameters. LNG produced for WBX (western buyers) contracts requires specific higher LNG density and heating value, while the remaining Lean LNG contracts specifies lower LNG HHV and density. With the conventional operating mode, Badak LNG is required to produce a single quality LNG with WBX specification, including for Lean contract. This mode operation requires large amount of LPG to be reinjected into LNG product for all operating Trains for both WBX and Lean LNG. Meanwhile LNG for Lean contract doesn't need LPG for reinjection. Therefore, currently Badak LNG implementing segregation mode (WBX and Lean production lines) which called dual HHV mode.

Dual HHV mode has been implemented since November 2017 by dividing Process trains into two types which are Rich Train and Lean Train. Rich LNG for WBX contract (HHV $\geq$ 1,107 btu/scf) is produced by Rich Trains, while lean LNG (HHV $<$ 1,107 btu/scf) is produced by Lean Trains. Meanwhile, rundown line systems, LNG Tanks, and Loading Docks are also divided to two types which contain rich and lean LNG product respectively.

Operational Optimizations

1. Optimization in Rich Train

a. Increasing Temperature inlet of Scrub Column

Actually, LPG consumption is used to increase HHV of natural gas outlet Scrub Column of Rich Train to required LNG product HHV. Higher HHV of natural gas outlet Scrub Column requires lower LPG for reinjection. Increasing HHV of overhead Scrub Column product can be conducted by increasing Temperature inlet of Scrub Column. Higher inlet feed temperature allows more C_2+ component in Top Column. Increasing 1°C inlet temperature increase HHV top product about 0.55 btu/scf. The correlation of Scrub Column inlet temperature and HHV of Scrub Column product has been defined by Plant test and the result can be seen on figure below.

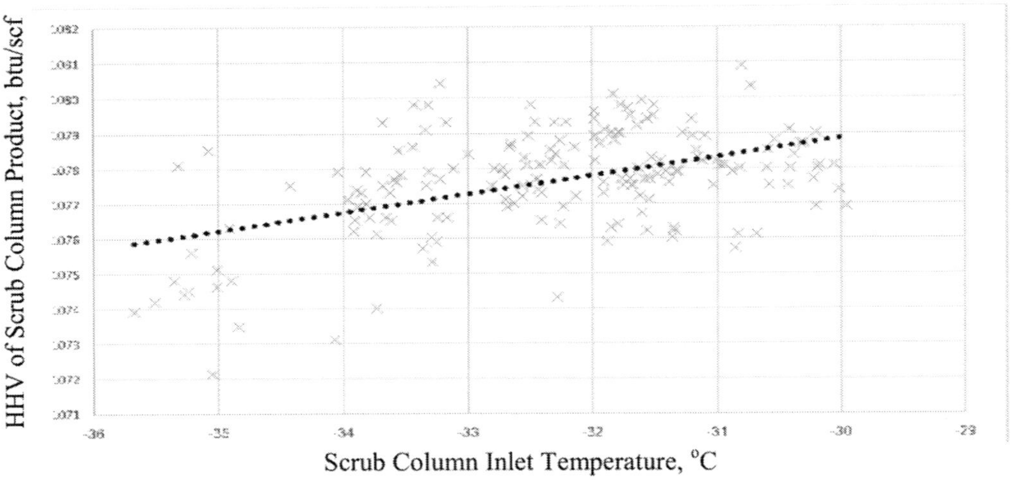

Figure 2. HHV of Scrub Column Product Profile

b. C_3 Interconnection between Process Train

Badak LNG have 5 LPG Tanks. Two of them is used to store LPG propane product, while the others is used to store LPG butane product. Since currently LPG propane Tanks contains contaminated propane that can causing freezing in Main Cryogenic Heat Exchanger if it used for reinjection, propane product from Process Train cannot be transferred to LPG propane Tanks. For Rich Train, propane product is reinjected to LNG product to increase HHV. Meanwhile for Lean Train, propane product is sent to refrigerant storages for refrigerant make up purpose, if refrigerant storages already full, it is reinjected to LNG product even it not necessary (there is no LNG HHV target of Lean Train).

One effort to reduce LPG external consumption is by transferring C_3 product from Lean Train to Rich Train to increase LNG's HHV. C_3 can be transferred from Lean Train trough existing C_3 Interconnecting line that can be seen on **Figure 3**.

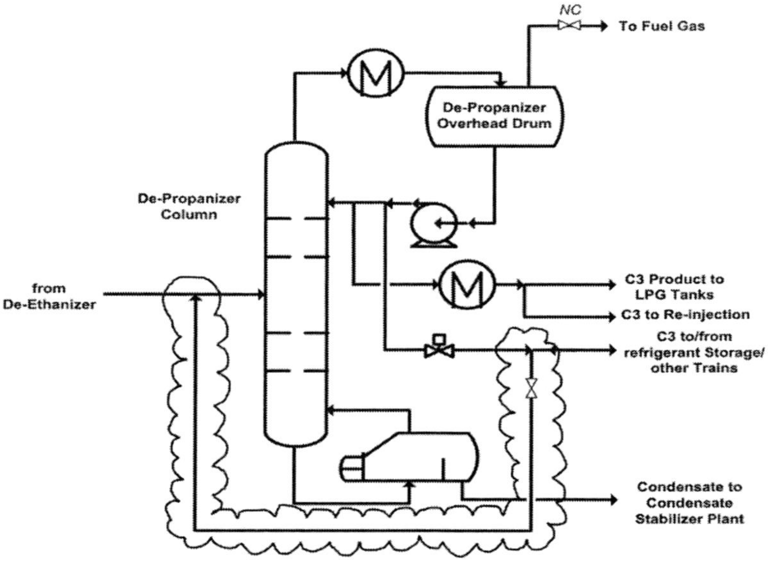

Figure 3. C3 Interconnecting Line Scheme

By implementing C_3 interconnecting, C_3 product in Rich Train can be increased for reinjection purposed. C_3 reinjection profile of Rich Train before and after C_3 Interconnecting implementation can be seen on **Figure 4**.

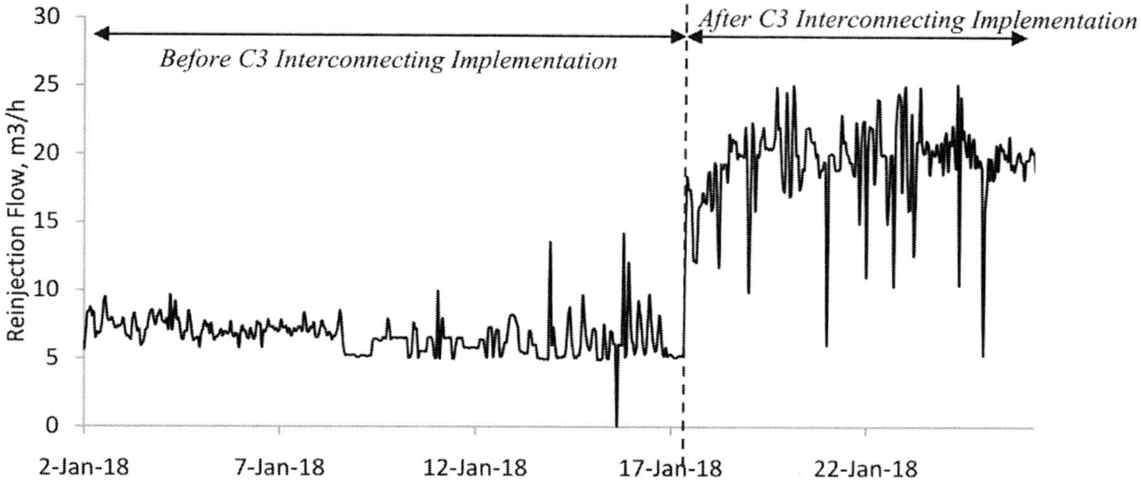

Figure 4. C_3 Reinjection of Rich Train

Figure-2 shows that by implementing C_3 interconnection, propane reinjection flow of Rich Train increased from 7 m^3/h to 20 m3/h. Therefore, additional external LPG consumption to achieve target of HHV is significantly decreased. If heating value of 1 m^3 butane equal with 1.24 m^3 propane, then implementing C_3 interconnection reduced additional C_4 consumption approximately 10.5 m^3/h. **Figure 5** below shows the LNG's HHV on Train E, G, and H after implementing C_3 interconnection without additional LPG reinjection.

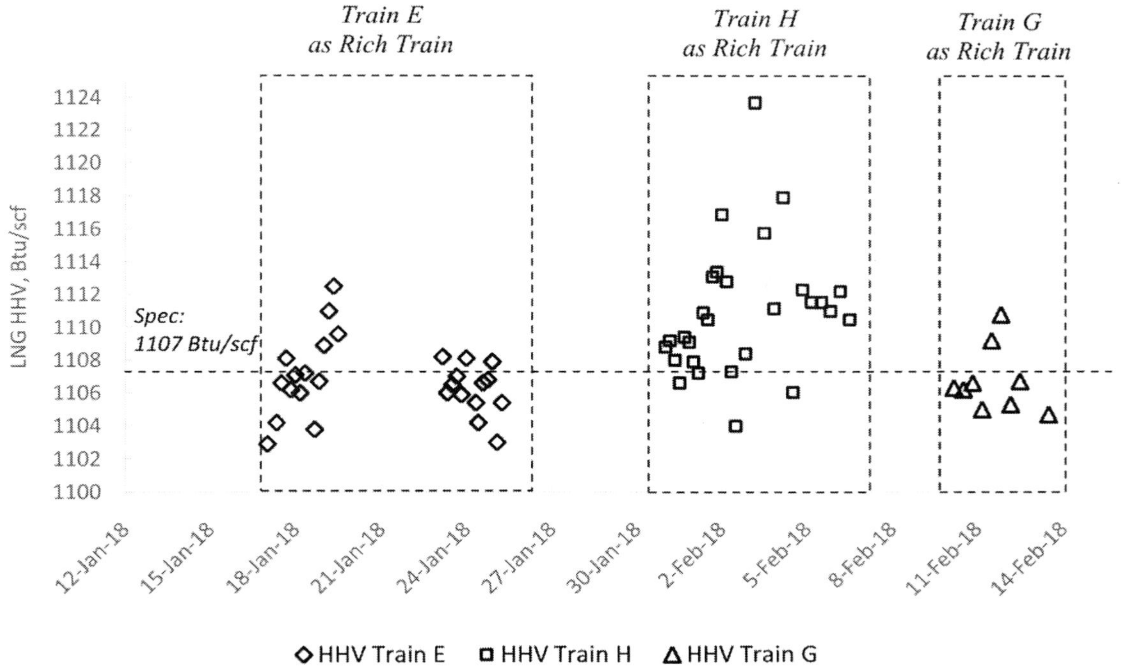

Figure 5. LNG's HHV of Rich Train during C_3 Interconnecting Implementation

c. Warmer LNG Product

Other effort that can be conducted to increase LNG product HHV is by increasing LNG product temperature. Increasing LNG product temperature can increase boiled of gas generation due to flashing. More light component flashing causes increasing HHV of remaining LNG. The correlation of LNG product temperature and HHV has been observed during process train produce lean LNG that can be seen on **Figure 6**. Based on observation, increasing 1°C LNG temperature, HHV increase 1 btu/scf in average. This option can be used to reduce LPG reinjection consumption to achieved target of LNG HHV.

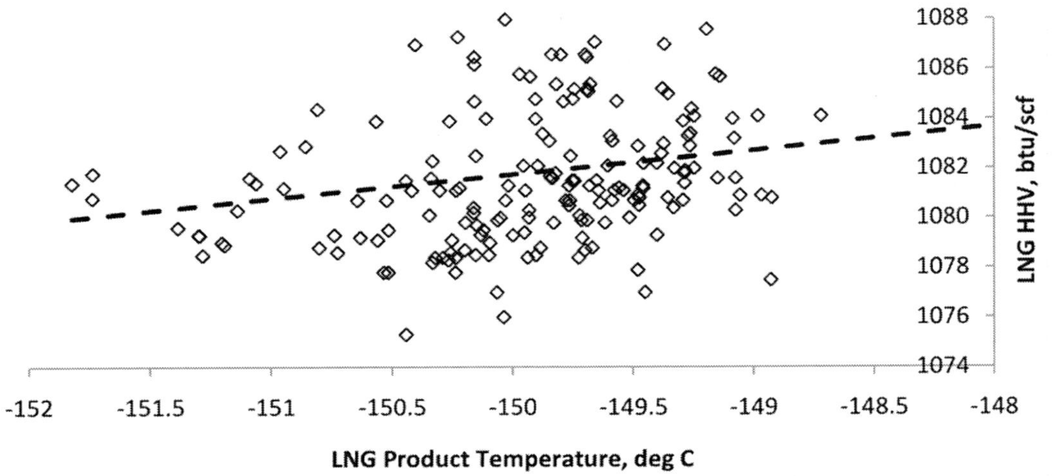

Figure 6. HHV Profile in Various LNG Temperature

2. Optimization in Lean Train

a. Decreasing Temperature inlet of Scrub Column

The opposite of optimization in Rich Train by increasing Scrub Column inlet temperature, Lean Train is optimized by decreasing inlet Scrub Column temperature. to enhance propane and butane production. Higher inlet feed temperature allows more C2+ component in bottom product of Column.

According to Plant test, the decreasing inlet Scrub Column from -30 °C to -35°C could effectively decrease C_3 content in Scrub Column top product from 1.8%-mol to 1.6% mol as shown in **Figure 7**. This decrement was equal with additional C_3 production of 79.5 m^3/day (3.3 m^3/h). This additional C_3 production could be transferred to Rich Train to decrease additional external LPG consumption to increase HHV.

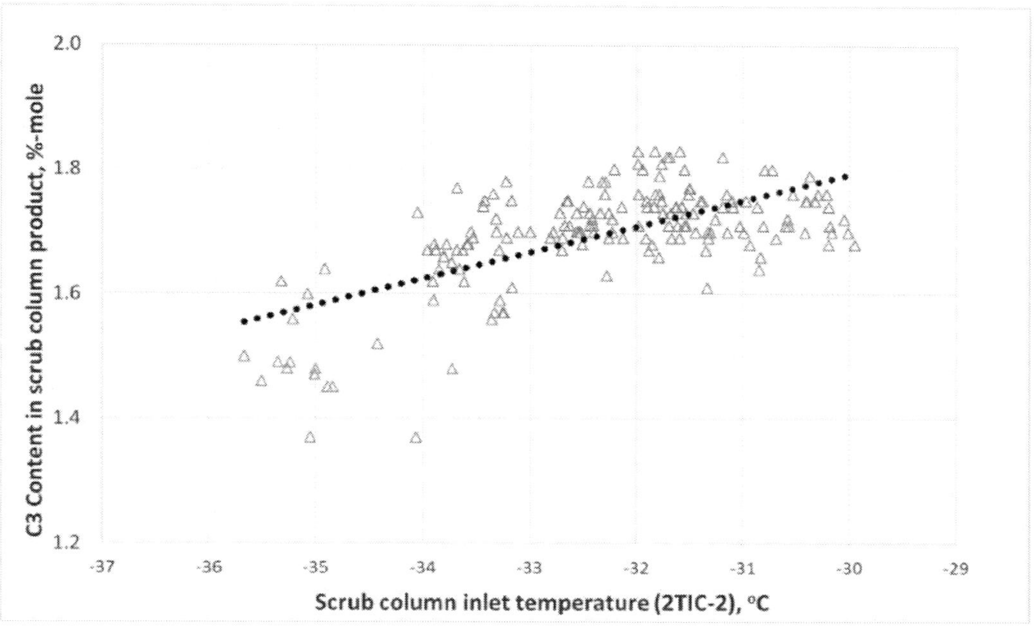

Figure 7. Total C_3 content in Scrub Column Product

According to the plant test, significant amount of C_2 were also successfully reduced from 2.9% mole to 2.7% mole in Scrub Column top product. This reduction is indirect parameter to indicate more C_2 flows to De-ethanizer to increase extraction.

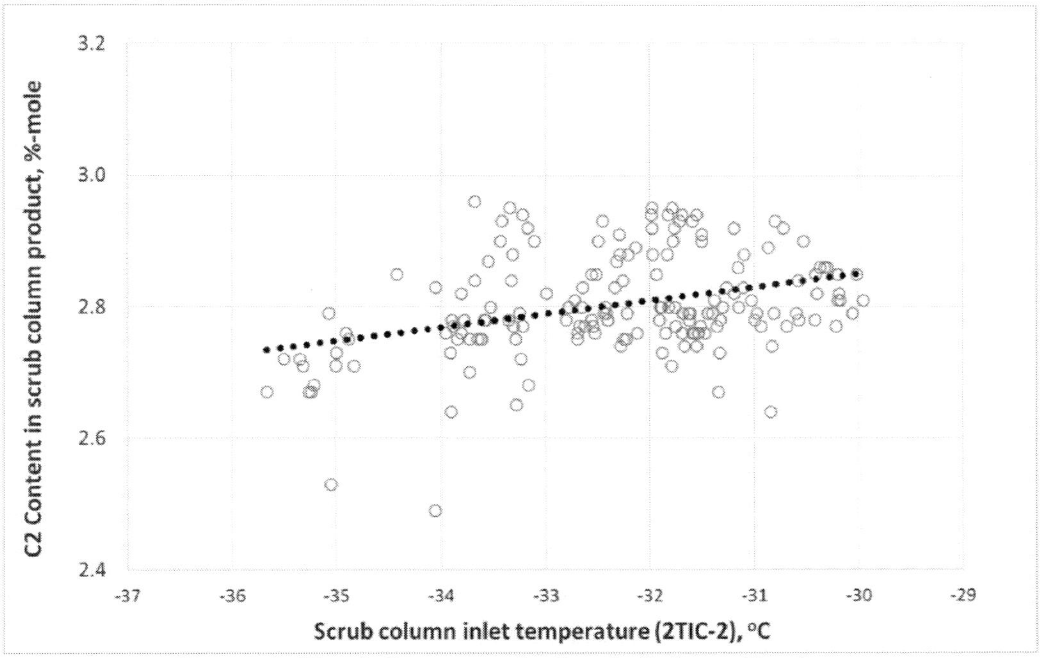

Figure 8. Total C_2 content in Scrub Column Product

Unfortunately, based on Plant test, the colder inlet Scrub Column increase C_4 content in Scrub Column top product. It is because decreasing inlet temperature of Scrub Column allows more C2+ component goes to bottom of Column and C_4 recycle requirement from Debutanizer will increase to maintain reflux flow as shown in **Figure 9**. The consequence is more C_4 carried over in Scrub Column top product as shown in **Figure 10.** According to the plant test, decreasing inlet Scrub Column temperature from -30 °C to -35.5 °C reduced C_4 production of 22.3 m3/day (0.93 m^3/h).

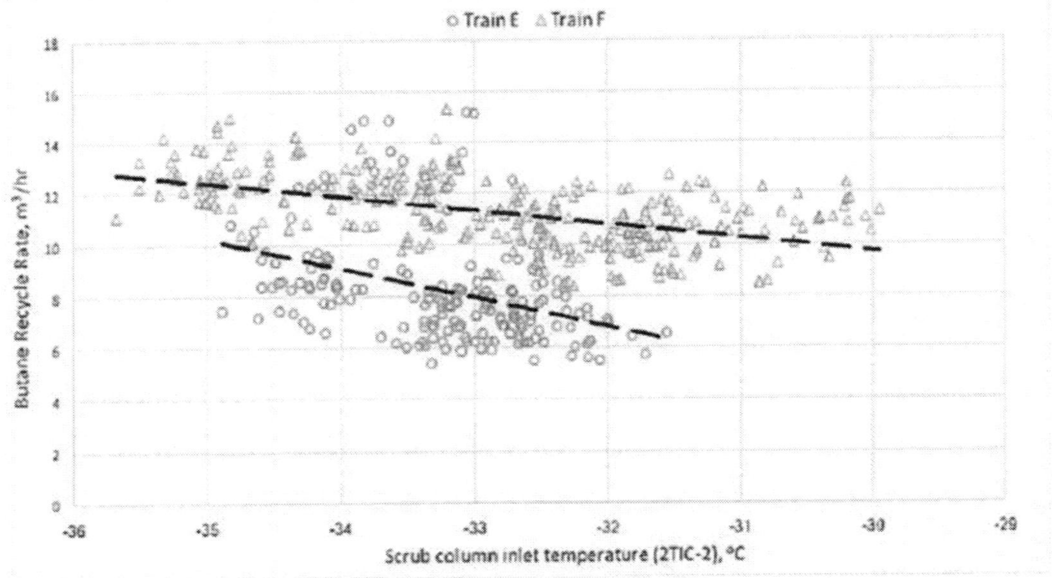

Figure 9. Butane Recycle Requirement Profile

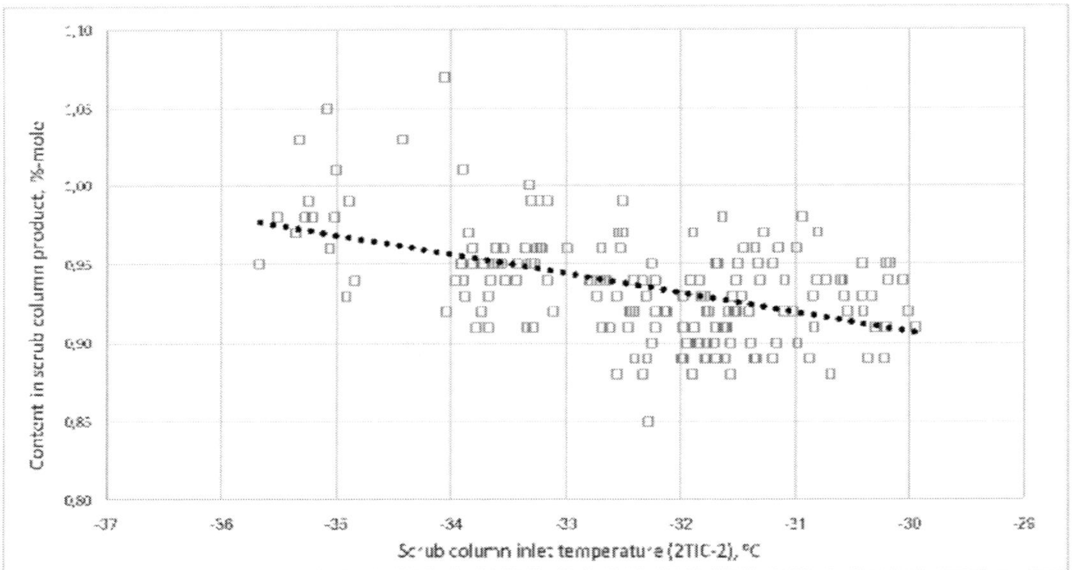

Figure 10. Total C_4 content in Scrub Column Product

b. Increasing Temperature of Debutanizer

Actually, C_4 product from Overhead Debutanizer has decreased due to leaner composition of feed gas. Moreover Debutanizer column stability has disturbed due to decreasing turn down ratio, indicated by low level of Overhead Debutanizer Drum. This problem can be solved by injecting external butane to Scrub Column through butane recycle line.

In order to minimized external butane consumption, Debutanizer top product also has been enhanced by increasing Debutanizer Reboiler Temperature. Increasing Debutanizer Reboiler temperature causing C5+ content in top product increase as shown on **Figure 11**. Therefore, butane product flow at top Debutanizer also increase, while bottom Debutanizer flow decrease as shown on **Figure 12**. Since there is now flow indicator installed at external butane stream, amount of butane consumption cannot be exactly observed. However, it can be determined indirectly from decreasing of bottom product. Amount of bottom product reduction is assumed equal with increasing amount of top product, thus it can be assumed equal with amount of C_4 external consumption reduction.

Based on observation, increasing Debutanizer Reboiler temperature 6 °C reduced bottom product flow of 1 m^3/h, thus, top product was increased approximately 1 m^3/h (external C_4 consumption was decreased approximately 1 m^3/h).

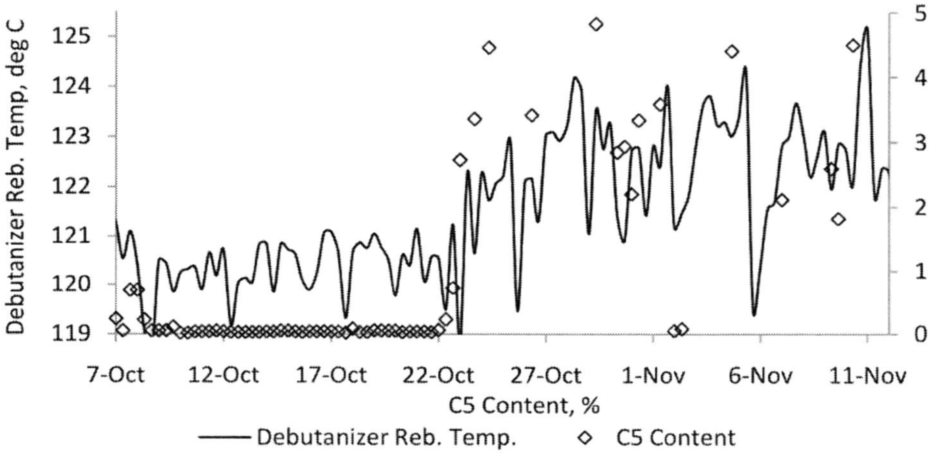

Figure 11. C5 Content Profile at Top Debutanizer

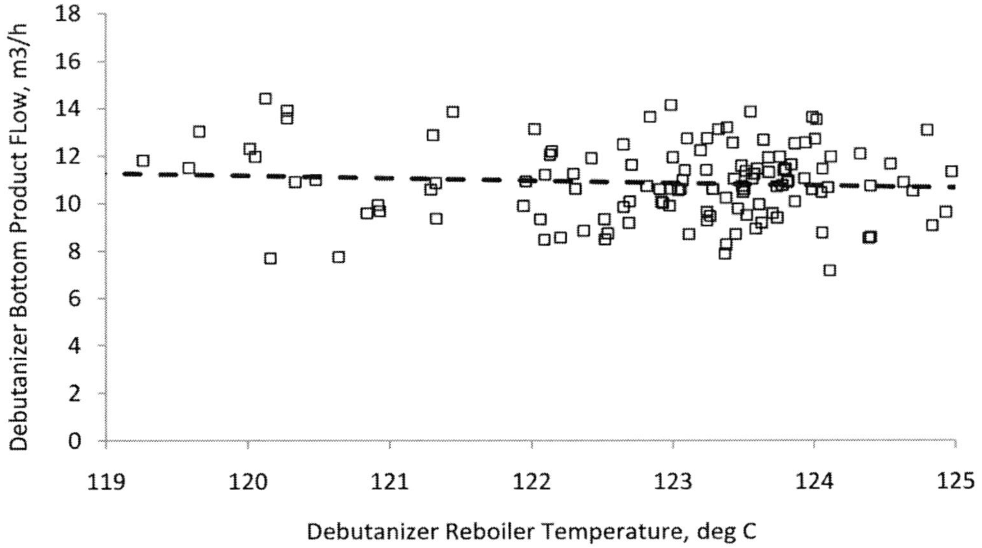

Figure 12. Bottom Debutanizer Flow Profile

c. Decreasing Temperature of Depropanizer

C_4 production also can be enhanced by decreasing Depropanizer Reboiler Temperature. Increasing Depropanizer Reboiler temperature causing bottom Depropanizer flow increase and butane product flow at top Debutanizer also increase. Based on observation, decreasing Depropanizer Reboiler temperature 10 °C increased bottom product flow of 5 m^3/h, thus, top product flow was increased approximately 5 m^3/h. By implementing the optimizations external C_4 consumption for column stabilization purpose can be reduced. Moreover, C_4 can be produced and stored in LPG Tanks.

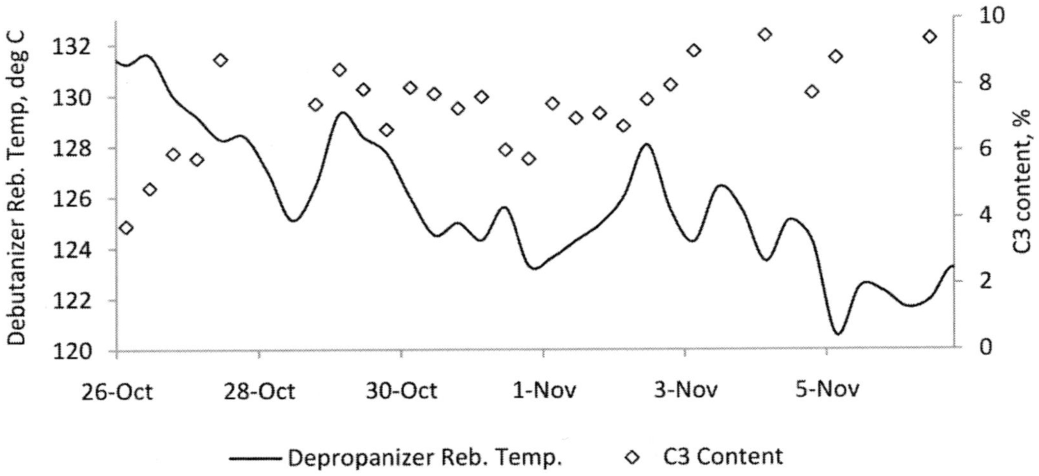

Figure 13. C3 Content Profile at Top Debutanizer

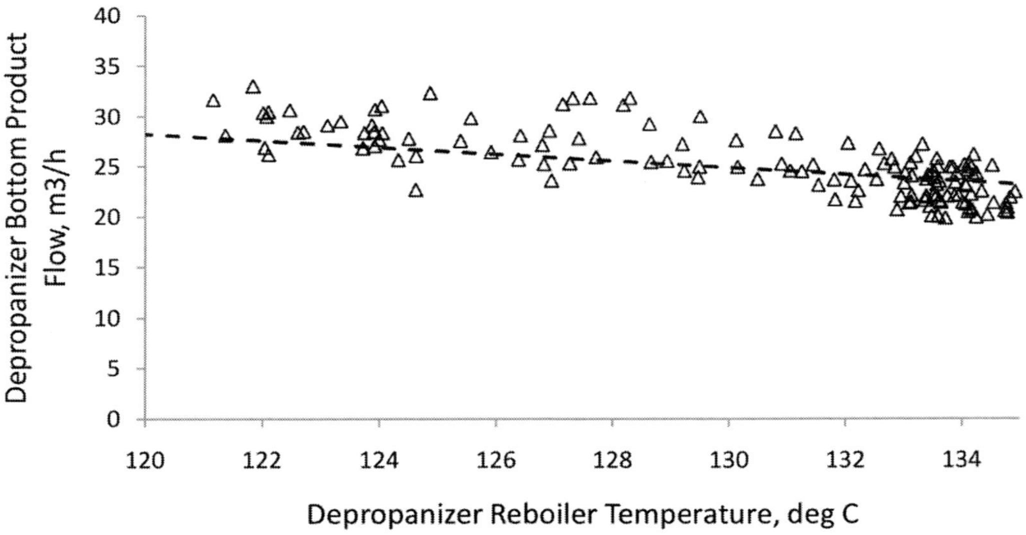

Figure 14. Bottom Depropanizer Flow Profile

3. Cost Saving

37,000 m^3 LPG butane was imported on January 2018 for reinjection purpose and was estimated run out at June 2018. With several optimizations mentioned above, LPG inventory was predicted still available until mid of February 2019. Actually, around 40,364 m^3 LPG butane can be saved in a year. If LPG butane price is 720 USD/MT, density of LPG butane is 596 kg/m^3, then LPG saving is approximately 17,321,000 USD/year.

4. Conclusions

1. Several optimizations that have been implemented on Rich Trains, which are:
    a. Increasing inlet temperature from -35°C to -30°C to increase Scrub Column Overhead HHV
    b. Implementing C3 interconnection to increased propane reinjection flow of Rich Train.
    c. Increasing LNG temperature to increased HHV whenever HHV still below the target of 1,107 btu/scf

2. The results of several optimizations that has been implemented on Lean Trains are:
    a. Decreasing inlet Scrub Column from -30 °C to -35°C to increase C3 production.
    b. Increasing Debutanizer Reboiler temperature until C5 content approaches 5% as per specifications.
    c. Decreasing Depropanizer Reboiler temperature until C3 content approaches 10% as per specifications.

3. By implementing optimizations on Rich Trains and Lean Trains during leaner feed gas entrance, Badak LNG could minimize LPG butane consumption up to 40,364 m^3/year equal with approximately 17,321,000 USD/year.

5. References

1. Darmawan, Robby Sukma (2018), "Maximize Extraction of NGL Product by Colder Scrub Column Operation," *PT Badak NGL,* Bontang.
2. Putra, Fajar Singgih Kurnia (2018), "Evaluation of C3 Interconnecting Implementation," *PT Badak NGL,* Bontang.
3. Setiawan, M. Arief (2018), "Badak LNG Dual Operation Mode Evaluation," *PT Badak NGL,* Bontang.

548928 Liquefaction Technology Selection for Mid-Scale LNG Plants

Liquefaction Technology Selection for Mid-scale LNG Plants

Saeid Mokhatab
Gas Processing Consultant, Halifax, NS, Canada
s.mokhatab@yahoo.com
Yoshitsugi Kikkawa,
Kikkawa Giken, Yokohama Japan,
yoshikikkawa@yahoo.co.jp

Abstract

Mid-scale liquefied natural gas (LNG) projects are forecasted to be the preferred solution for niche LNG markets as well as for the monetization of smaller gas volumes at reduced investment cost. As such, there is great interest in developing successful solutions for mid-scale LNG projects that may have advantages over competing baseload projects. The key step in addressing an economical solution to the mid-scale LNG project development is selecting a less complex and lower cost liquefaction technology, which best fits the size range of the project. Mid-scale LNG plants may be located onshore or offshore. Although technical requirements are different in each case, simplicity and modularization are main targets in both. This paper discusses various natural gas liquefaction technologies to allow a selection that best fits the needs of land-based mid-scale LNG applications.

549968 California Energy Commission Integrated Energy Policy Report

Every two years, the California Energy Commission prepares the Integrated Energy Policy Report (IEPR). The IEPR highlights the implementation of California's innovative policies and the role they have played in establishing a clean energy economy. The 2018 Update report details several key energy issues and will encompass new analyses.

551808 Effects of Elevated CO_2 Concentrations on Yields of Agricultural Crops

That elevated concentrations of CO_2 can increase plant growth has been known since at least 1804, when a scientist named De Saussure reported that peas exposed to high CO_2 concentrations grew better than control plants in ambient air (about 300 ppm at that time). Since then, thousands of experiments with elevated concentrations have been conducted, mostly in greenhouses and other types of enclosures, and almost of them similarly showed increases in growth and/or yield. Indeed, it is standard horticultural practice to enrich greenhouses to about 1000 ppm CO_2 when conditions are cool enough that so that the greenhouses do not have to be ventilated for temperature control. Then, when it was realized that the atmospheric CO_2 concentrations were increasing, scientists wondered what this portended for future plant growth and agricultural productivity in open fields. Therefore, about 30 years ago, free-air CO_2 enrichment (FACE) technology was developed that enabled the air above open-field plots to be enriched with CO_2 for entire growing seasons. Since then, FACE experiments have been conducted on cotton, wheat, ryegrass, clover, potato, grape, rice, barley, sugar beet, soybean, cassava, rape, mustard, coffee (C_3 crops), and sorghum and maize (C_4 crops). Elevated CO_2 (550 ppm from an ambient concentration of about 353 ppm in 1990) decreased evapotranspiration about 10% on average and increased canopy temperatures about 0.7°C. Biomass and yield were increased by FACE in all C_3 species, but not in C_4 species except when water was limiting. Yields of C_3 grain crops were increased on average about 19%, while other C_3 crops were increased even more.

550130 The Development of the Detal Process

This presentation will present the highlights of the development of the Detal process and the impact of Bipin on the development and commercialization. The Detal process is the Solid Bed Alkylation process that replaced HF Alkylation in the UOP Detergent complex.

556730 The Origin, Development and Commercialization of the UOP Oleflex™ Process, Bipin Vora's Vision and Leadership in Making It Happen

Since its commercialization in 1990, the UOP Oleflex™ process has become the world's dominant technology for on-purpose olefin production from LPG. The Oleflex process accounts for more than 30 million metric tons/year of high purity olefin products – Propylene, Isobutylene and linear C4 olefins for a wide variety of petrochemical and fuel applications. More than 30 Oleflex units are on-stream worldwide, and more than 70 Oleflex units been licensed, designed and in construction. This technology continues to advance to ever larger capacity plants with increasingly attractive CAPEX, OPEX and yields.

Bipin Vora's work at UOP as a visionary, inventor, and promotor of on-purpose olefin production has had an important and lasting influence on the Oleflex process. This presentation covers the origins of the Oleflex process and focuses on Bipin's key contributions that led to this technology becoming the industry standard for on-purpose light olefin production.

550722 The Research, Development and Commercialization of the Methanol-to-Olefins (MTO) Process

The last great technology in Bipin Vora's long career in Process Design and Development was the Methanol to Olefins process. The talk will discuss the research and development that lead to the commercialization of MTO

550691 The Development and Commercialization of the Oleflex Process Unit. - This Talk Will Discuss the Role Bipin Vora Played in the Successful Deployment of the Olflex (C3 and C4 Olefin Process) Process

UOP is a company that has commercialized one new process technology per year for 105 years. It has taken unwavering management commitment to identify new opportunities, process risk, provide resources, and most of all to support and develop multi-skilled professionals as epitomized by this summary of the career of Bipin Vora. This paper focuses on his development of a process for converting propane to propylene, although he has played a key role in other applications such as biodegradable detergents and conversion of methane to olefins.

Licensing companies strive to develop processes that have broad appeal and generate numerous commercial opportunities. When successful commercial operation has been demonstrated subsequent units can be designed quickly. While Bipin's successes have been built many times, his expertise is in the space of blank paper and a market need. He starts with physical data about potential feedstocks and products, evaluates reaction pathways, works with research personnel to find a suitable catalyst, oversees pilot plant demonstration of catalyst performance and life, evaluates reactor options, prepares a process flow diagram that integrates reaction and separation sections, studies potential challenges such as thermal cracking, and then works with process engineers to prepare a Piping and Instrument diagram detailing every line and valve in the plant and a Schedule A book specifying details of every vessel, pump, compressor, valve, fired heater, and heat exchanger. Before a new process is ready for licensing it may be necessary to do computer and physical modeling especially of reactor details to determine appropriate scaleup parameters. Note that Bipin's close involvement with all phases of process development over time periods as long at ten years means nothing is lost due to incomplete handoffs to new personnel.

Selling a new process license is challenging, as most customers like to visit an operating commercial unit before committing funds to their own unit. Bipin is quite effective in this stage of demonstrating analogies with other processes and detailing the investigations that give him confidence to stand behind performance guarantees. When the Oleflex design was turned into construction drawings Bipin noticed a long high temperature line behind a heater and a reactor that had a residence time much higher than his lab studies had shown to be acceptable. He worked with a highly experienced process designer from the engineering department who was willing to step away from "the way we have always done it" to radically change the reactor/heater layout to comply with the residence time criterion.

Bipin's career path prepared him to excel in all aspects of developing a new process. He spent time as a hands-on operator of pilot plants, supervised the operation and data analysis from other pilot plants, started up several process units around the world, worked in process design development, managed a number of research and development projects, provided technical support to marketing personnel, became expert in patents, and mastered strategic planning for new process opportunities.

The culture of UOP encouraged its people to keep looking for improvements to existing technology, to take chances with big ideas, and to collaborate to make things happen. This starts with recruiting and developing the best people available. Resources were available to chase new ideas. When you try many new things not everything will work out as hoped but failure was treated as a learning opportunity. Rotation among assignments is another key. Managers do not hang on to promising employees, rather they find new challenges outside their domain and in turn bring in high caliber replacements. This encourages communication and cooperation across departmental lines.

How can young people emulate Bipin and be amazingly effective chemical engineers? It is not something that can be taught in a university. But educators can encourage their students to be curious and creative, to think beyond purely chemical engineering concepts, to seek opportunities to get their hands dirty working with operation and maintenance personnel a few decades older than they are. Student chapters of AIChE can expose students to talks by experienced engineers and organize plant trips to see what the real world looks like. And students can be encouraged to seek jobs with the potential exposure to every business function to build personal credibility on paths to either higher level management or technical positions.

Sensitivity Analysis of Transport Conditions on Liquified Gas Hazards

Ryan Hart, Ph.D., P.E.
Jessica M. Morris, Ph.D.
Delmar "Trey" Morrison, Ph.D., P.E.
Exponent, Inc.
4580 Weaver Parkway, Suite 100,
Warrenville, Illinois, USA 60555
jmorris@exponent.com

Prepared for Presentation at
American Institute of Chemical Engineers
2019 Spring Meeting and 15th Global Congress on Process Safety
New Orleans, LA
March 31–April 3, 2019

Sensitivity Analysis of Transport Conditions on Liquified Gas Hazards

Ryan Hart, Ph.D., P.E.
Jessica M. Morris, Ph.D.
Delmar "Trey" Morrison, Ph.D., P.E.
Exponent, Inc.
4580 Weaver Parkway, Suite 100,
Warrenville, Illinois, USA 60555
jmorris@exponent.com

Keywords: Consequence Modeling, Hazard Identification and Risk Analysis, LNG

Abstract

Over the last decade, regulatory and societal scrutiny of the hazards of transportation of flammable liquids and liquefied gases has risen considerably. Currently, some of these materials are routinely transported in bulk over the road, rail, and marine, but others, namely liquefied natural gas (LNG), are not. An integral aspect of demonstrating the safety of these activities is the application of risk assessment to the transportation activities. Risk assessments for liquefied gas (pressurized liquefied gases and cryogenic liquefied gases) transportation involve calculating hazardous consequences from loss of containment events due to accidental releases. In typical hazard/risk consequence calculation methods, the pressure and temperature used to calculate the hazards are those values expected under the normal operation of a steady state process. For non-steady state processes, such as the lengthy transportation of cryogenic materials or vessel rupture due to mechanical impact, the properties of the fluid inside the vessel can be significantly different from the expected or normal condition. The properties of the release may have a significant impact on the hazard calculations and resulting risk profile. For example, a cryogenic flammable liquid stored at its normal boiling point will likely result in liquid pool formation upon release—exhibiting flash fire, explosion, or pool fire hazards. However, if the same liquid is released either (1) under significant pressure due to mechanical impact or (2) at thermodynamic equilibrium at a temperature well beyond its normal boiling point, the release may not result in any liquid spillage—exhibiting fireball, flash fire, explosion, and jet fire hazards. In this paper we explore how these possible changes in the release conditions can affect both the hazard consequence calculations and risk assessments for the transportation of these materials. The materials explored shall include liquefied natural gas and liquefied petroleum gas.

1 Introduction

With gasses having a volume ~600x greater than liquids, it is more economic to transport some chemicals in their liquid state. Flammable chemicals such as liquified natural gas (LNG), liquified petroleum gas (LPG), and ethylene are commonly stored and transported as a liquid. Transitioning from a gas to a liquid is easily achievable through changing the chemical's temperature and/or pressure. LNG and ethylene are stored at atmospheric pressure at -161°C and -103.9°C, respectively, where propane is stored at 106 psig and 20°C.

1.1 Transportation of Flammable Liquefied Gases

While many flammable gases are transported as liquids or gases by vessel or pipeline, there are not enough waterways or pipelines to transport all flammable gases to their final destination. Flammable gases are commonly liquefied then transported via specialized containers through railway or road networks. According to 49 CFR §173.315, pressurized liquefied petroleum gases (LPGs) are shipped via rail in DOT-51 single-walled, pressurized tank cars while cryogenic liquefied gases (e.g., LNG, ethylene, and hydrogen) are shipped via rail in DOT-113 double-walled, pressurized tank cars according to 49 CFR §173.319. For trucking, similar containers to these are used: M-331trailers and MC-338 trailers for pressurized liquefied gases and cryogenic liquefied gases, respectively. A third mode of transportation uses portable cargo containers with ISO specifications such as T-50 and T-75 for pressurized liquefied gases and cryogenic liquefied gases, respectively.

The transportation method of rail or road provides additional hazards specific to these systems. Accidents on both rail and road can occur through collision, fire, derailment, blockage, or natural disaster, among others.[1] Depending on the accident dynamics, the flammable liquefied gas may be released at non-equilibrium conditions. Due to changes in temperature and pressure, the accident occurring plays a large effect on release dynamics of the material. Rarely do accidents occur under steady-state conditions. For cargo tanks involved in an accident after a lengthy transportation period or involving mechanical impact, such as derailment, the properties of the fluid can vary greatly from steady-state or normal conditions. An example of the quantitative release from a multiple tank car derailment in Lac Megantic is shown below in Figure 1 to illustrate that the damage to individual cars in a catastrophic derailment may vary.

Figure 1. Petroleum released from a tank derailment in Quebec resulting in 47 fatalities from the fires and explosions. Colored tanks relative to release size: orange–large, yellow–medium, and blue–small.[2]

From Figure 1, the more deformed tanks had a larger immediate release into the environment. Releases under non-equilibrium pressure and/or temperature conditions affect the release dynamics, resulting in different predictions for consequence and risk analysis than would have been performed under normal operating conditions.

1.2 Consequence and Risk

Risk is a function of the (1) estimated frequency of the event, (2) estimated consequences, and (3) hypothetical scenario. Determining risk can be achieved from understanding the system, potential hazards, incidents, and estimating the consequences.[3] Combining consequences and frequencies determines the qualitative risk for the system. In approximately 95.5% of pressurized railcar incidents, there is no loss of containment.[4] Although the probability is low for loss of containment, the consequences of the release conditions analyzed result in different levels of risk. Risk is directly proportional to consequence. For this study, the estimated consequences, and therefore risk, from hypothetical release scenarios will be investigated, specifically the effect of temperature and pressure on release dynamics of flammable liquids transported via rail or road.

Current risk assessments for accidental release of flammable liquefied gases evaluate the incident under steady-state temperature and pressure. While this approach is generally accepted, loss of contaminant can fall under non-steady state conditions if released after a lengthy transport, under mechanical impact, or other incidents. Appropriate consideration of potential temperature and pressure deviations from steady-state observed in loss of contaminant incidents can affect the overall hazardous consequences. Therefore, it is of utmost importance to take into consideration the temperature and pressure differences when performing the consequence analysis for release of flammable liquefied gases.

A thorough understanding of the effects of temperature and pressure on release dynamics of flammable liquefied gases will result in a more inclusive consequence analysis that can occur with the accidental release conditions of flammable liquefied gases during transportation. This paper focuses on the temperature and pressure fluctuations that can occur with the accidental release of flammable liquefied gases during transportation for a better understanding of the potential consequences due to the varying release dynamics observed during the transportation stage due to temperature and pressure.

2 Analytical Program

To determine the impact of temperature and pressure on the potential consequences, an extensive analysis matrix was created. The analytical program included a series of commonly transported flammable liquefied gases, temperature and pressure scenarios reflective of realistic accidents, and a selection of leak scenarios.

2.1 Hazardous Chemicals

The three chemicals investigated in this study were LNG, LPG, and ethylene. For simplicity purposes, LNG and LPG were modeled as pure methane and propane, respectively. These three chemicals were chosen due to their flammable properties and frequent transportation in liquid state across both rail and road.

2.2 Temperatures and Pressures

Variations in temperature and pressure have a large effect on the state of release. All chemicals chosen for this study are commonly transported in liquid phase which may or may not be released as a liquid or gas depending on their properties when released into the atmosphere. Realistically, there are three cases in which flammable liquefied gases will be accidentally released from their storage container into the atmosphere.

1. Cold Case: loading conditions for the chemical at its natural boiling point and atmospheric pressure (LNG and ethylene) or slightly pressurized at room temperature (LPG).
2. Warm Case: increased temperature due to time in tank. Saturation temperature is observed at the pressure relief valve setpoint.
3. Accident Case: tank is compressed (likely to occur in an accident) resulting in the pressure relief valve setpoint and normal boiling point from recent loading.

The specific temperature and pressure combinations vary based on the chemical properties. Phase diagrams with the maximum allowable working pressure (MAWP) of the transportation containers, normal boiling points, and specified cases mentioned above are shown below as individual points for LNG, LPG, and ethylene.

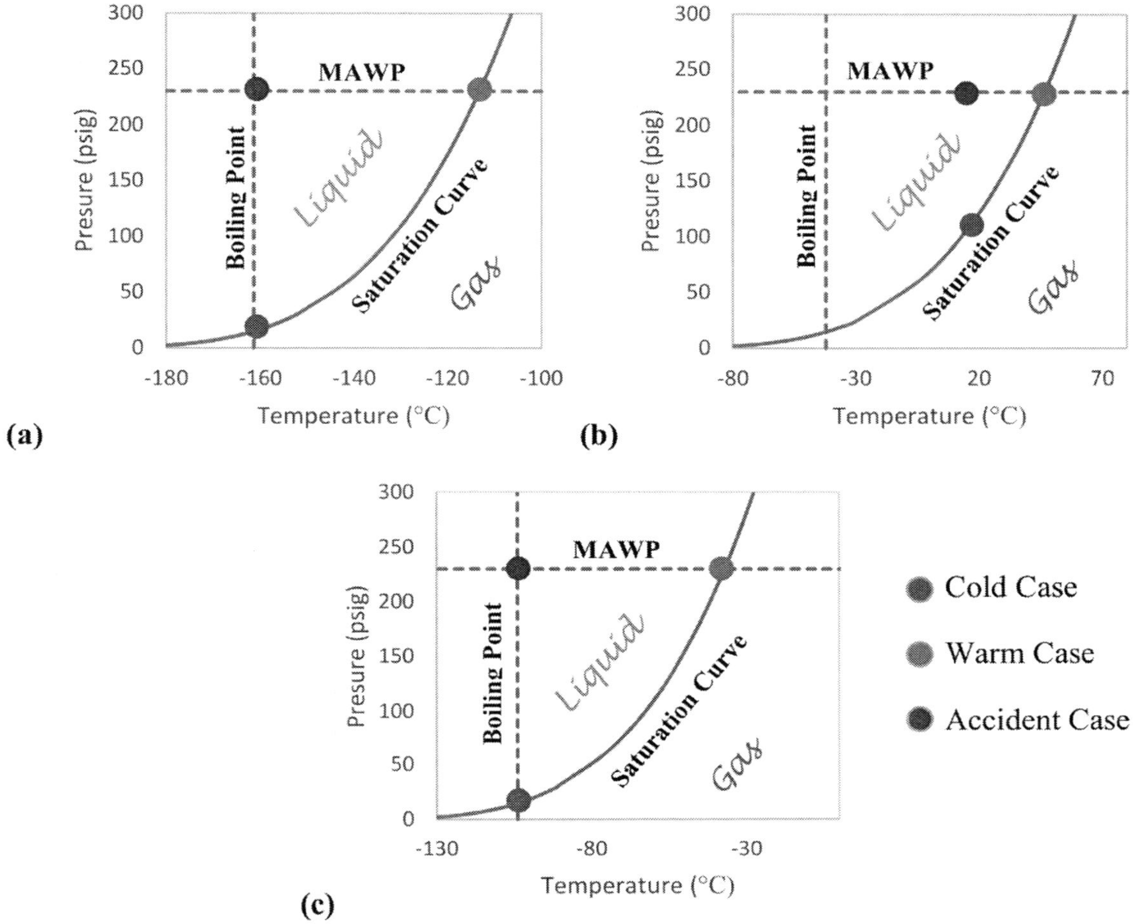

Figure 2. Phase diagrams for (a) LNG, (b) LPG, and (c) ethylene with cases

From the phase diagram, the temperature and pressure for each chemical are determined for each of the potential release cases. The specified temperatures and pressures are shown below in Table 1 for each release case.

Table 1. Temperature and pressure values for specified test cases for all chemicals

	LNG		LPG		Ethylene	
Cases	**T (°C)**	**P (psig)**	**T (°C)**	**P (psig)**	**T (C)**	**P (psig)**
Cold	-161.4	0.0	20.0	106.8	-103.8	0.0
Warm	-111.7	230.0	49.2	230.0	-34.8	230.0
Accident	-161.4	230.0	20.0	230.0	-103.8	230.0

The temperatures and pressures determined for the three realistic test cases are inclusive of the three scenarios in which accidents can occur for the flammable liquefied gases investigated.

2.3 *Leak Scenario*

Three leak scenarios were selected for modeling purposes to encompass a wide range of potential leak scenarios. The connection port for liquid in storage containers to be received, discharged, and recovered is typically 3" in diameter[5] and will be used as the first leak scenario. Other valve sizes for standardized transportation containers span from ½" to 1½" in diameter. A diameter of 1" was chosen as a median value to encompass the varying smaller pipe fittings as the second leak scenario. The third leak scenario was chosen as a catastrophic release to encompass the worst-case scenario as the most conservative option. Within the modeling software, a catastrophic release is defined as 100% of the chemical released instantaneously at the specified temperature and pressure.

2.4 *Analysis Matrix*

An overall matrix encompassing three commonly transported flammable liquefied gases, three potential leak scenarios, and three realistic temperature and pressure case combinations was developed. The different release dynamics and the consequences of each set of conditions will be investigated resulting in 27 overall test cases.

3 Modeling of Release Simulations

Simulations were performed to model the effects of temperature and pressure on commonly transported flammable liquefied gases. The modeling package used for these simulations was DNV GL Phast 8.11. Standards were chosen across a range of parameters to provide a base for the specified conditions investigated in this study. Standards regarding volume, MAWP, weather, wind speed, temperatures, and other conditions are discussed and specified below.

3.1 *Transport Containers*

For transportation on both rail and road, liquefied flammable gases are transported via specialized containers to keep the temperature and pressure consistent in addition to meeting the pressure and piping specifications for loading and unloading flammable liquefied gases. While no tank is 100% insulated, the International Organization for Standardization (ISO) containers for cryogenic liquids from some manufacturers maintain the boil-off-rate, due to increases in temperature and therefore pressure, at 0.15-0.3% per day.[6]

For the standardization of a tank size and pressure, ISO containers were chosen as the focus due to their use across both rail and road for transportation of specific flammable liquefied gases. Standard 20' ISO containers have a MAWP of 100, 150, 180, and 230 psig for compliance with ASME/DOT design. For conservative results, the largest MAWP was chosen at 230 psig. Standard 20' ISO containers for cryogenic bulk liquids can transport a maximum volume of 5,326 gallons. For simplicity, a standard volume of 5,000 gallons was chosen for the total mass in this study.

3.2 Modeling Parameters

Both the NFPA 59A regulations for siting requirements regarding flammable vapor-gas dispersion protection[7] and the 49 CFA transportation siting guidelines for liquefied natural gas[8] were used for conservative modeling input parameters for atmospheric conditions within the modeling software. The following dispersion parameters were used in computing dispersion distances for all chemicals investigated:

1. Pasquill Stability Class: F
2. Wind speed: 2 m/s (4.5 mph) at reference height of 10m
3. Relative humidity: 50%
4. Atmospheric temperature = average in the region (20°C / 68°F)

While these conditions are required for modeling LNG, these specifications provided the most conservative test conditions therefore used across all flammable liquefied gases modeled in these simulations. All default settings were untouched other than what is specified above, chemical, temperature, pressure, leak scenario, and a 10" head in the tank to serve as a driving force for the releases at atmospheric pressure (cold case). The 10" head was applied to all case scenarios for standardization across all simulations.

Default conditions within Phast 8.11 position the orientation of the discharge horizontally. Additional simulations were run for one test case at various discharge orientations: vertical, down, horizontal, and angled from horizontal to determine the impact on consequence due to discharge orientation. A discharge orientation straight down resulted in errors for the jet fire calculations and increased pooling, and therefore large, unbounded pools. A vertical discharge orientation resulted in vertical jet flames and a relatively small LFL distance measured at ground level due to reference resulting in unrealistic consequence. Angled from horizontal provided similar results and consequence to that of horizontal discharge orientation. The default setting of horizontal discharge was used throughout the simulations investigated here.

4 Results and Discussion

The full analytical program investigated three flammable liquefied gases, three leak scenarios, and three realistic test cases with specified temperatures and pressures for the individual case and chemical combination. Consequence as a result of release dynamics, rain out, LFL, flash fire, and jet fire are all a function of the test conditions presented and are discussed below.

4.1 Release Dynamics

For catastrophic releases, Phast 8.11 models the full release as an instantaneous release. This results in 100% of the chemical mass released instantaneously at the specified temperature and pressure. Due to the characteristics of instantaneous releases, the catastrophic test cases for release dynamics are not portrayed for release rate, duration, or mass. The 1" and 3" leaks for all flammable liquefied gases and case specifications are shown below.

Table 2. Release rates, durations, and total mass released for all flammable gases for 1" and 3" leak scenarios for all temperature and pressure cases combinations

	Release Rate [lb/min]			Release Duration [s]	
LNG	**1"**	**3"**		**1"**	**3"**
Cold Case	49.0	441		3,600	2,400
Warm Case	1,370	12,400		606	67.4
Accident Case	1,470	13,300		717	80
LPG	**1"**	**3"**		**1"**	**3"**
Cold Case	1,140	10,300		1,100	122
Warm Case	1,630	14,700		687	76.4
Accident Case	1,670	15,000		748	83.1
Ethylene	**1"**	**3"**		**1"**	**3"**
Cold Case	51.9	467		3,600	3,050
Warm Case	1,600	14,400		700	77.8
Accident Case	1,740	15,700		818	90.8

From Table 2, the accident case showed the largest release rate and total release mass across all chemicals and leak scenarios. While LNG and ethylene show similar trends for release rate and mass increasing from cold to warm to accident, LPG only demonstrates these trends with release rate. The overall release mass for both the cold case and accident case for LPG are equivalent and within ~11% of all LPG releases. All 1" leak scenarios result in larger release durations than the 3" leak scenario. Cold cases result in the longest release durations followed by warm cases and then accident cases across all flammable liquefied gases and leak scenarios.

4.2 Pooling and Flashing

As shown with the phase diagrams in Figure 1, the saturation curve provides the distinction between liquid or gas phase. With flammable liquefied gases stored at their saturation point or above, liquid releases are plausible. Pooling occurs when liquid accumulates under the release. Pool cases provide an additional dispersion mechanism due to the evaporation of the pool and additional consequence due to accumulation of a flammable liquid over a large surface area. Pooling can occur from a release if the output is liquid or of two-phase flow due to storage conditions. Flashing can occur when the pressure drops below the bubble point pressure, changing the phase of the output flow to liquid.[9] A bar graph of the pooling occurring for all flammable liquids, leak scenarios, and cases are shown below.

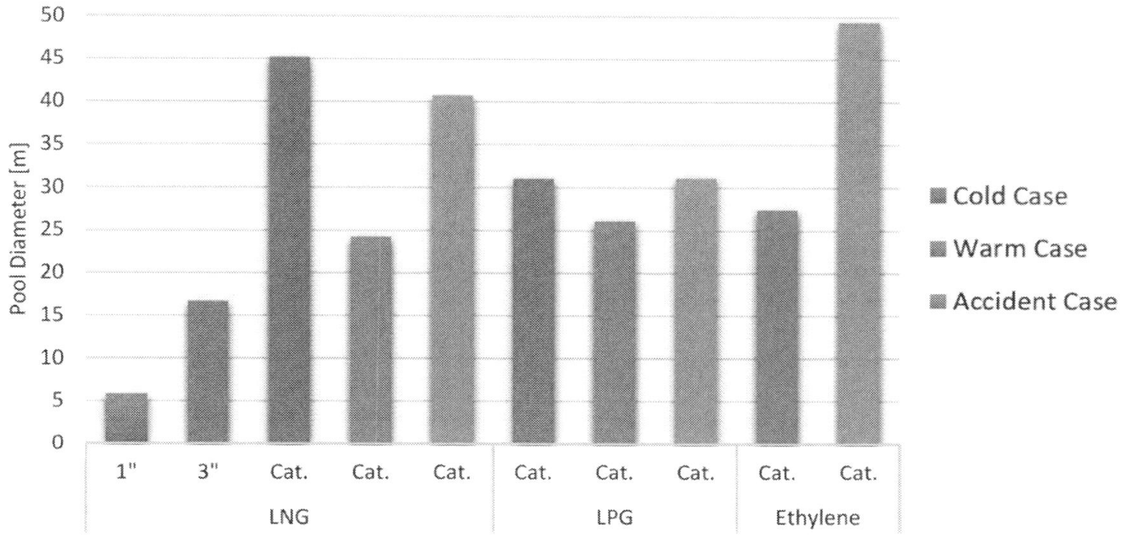

Figure 3. Pool diameters from simulations involving LNG, LPG, and ethylene for all leak scenarios and test cases

Almost all catastrophic releases result in pooling. All the LNG releases at loading conditions (cold case) result in pooling under the release. Liquid pools provide an additional hazard from potential pool fires. With regard to consequences of pooling, any catastrophic release shown above has the largest risk with respect to pooling followed by the 1" and 3" scenarios of LNG at loading conditions.

4.3 Flash Fires and Jet Fires

A flash fire results in the ignition of the vapor cloud around a release without an increase in the combustion rate.[10] If an increase in combustion rate occurs, an explosion takes place resulting in overpressure. For the simulations performed here, only jet fires and flash fires are reported for comparison of potential consequence and risk. The distances to the LFL and potential flash fires are reported for each simulation in Figure 4.

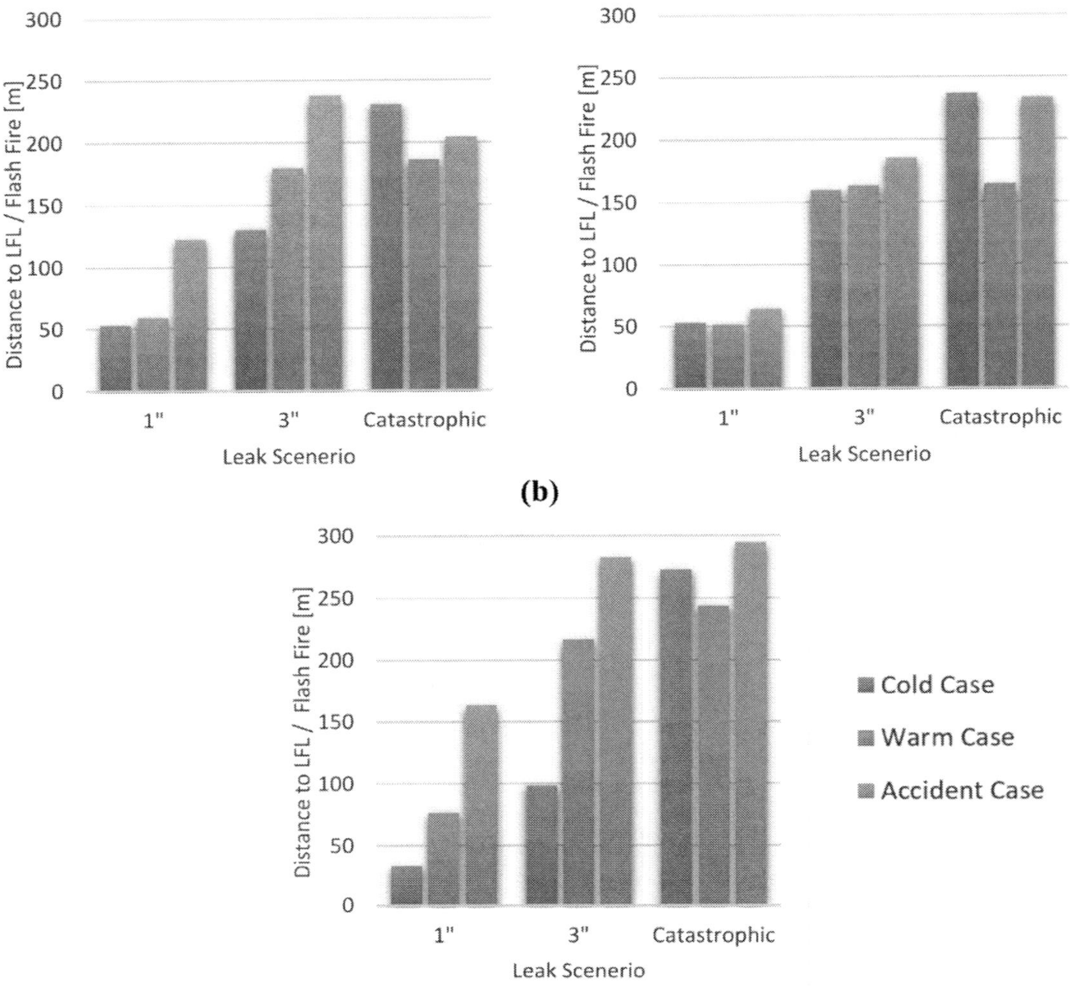

(a) (b)

(c)

Figure 4. Distances to LFL / Flash Fire for (a) LNG, (b) LPG, and (c) ethylene for all leak scenarios and test cases

From Figure 4, all simulations result in the release of flammable liquefied gases with a potential for flash fire. The 1" leak scenario for all flammable liquefied gases within the test cases results in smaller distances that exceed the LFL for a potential flash fire. The 3" leak scenario from the accident case results in the largest distance from the initial source for potential flash fire and an area that exceeds the LFL for LNG. All other 3" leak scenarios for all cases and chemicals have relatively moderate results. The catastrophic releases, designed to provide the most conservative results, show the largest LFL and flash fire diameters for flammable liquefied gases within the test cases. For the catastrophic releases for both LNG and LPG, a catastrophic release at loading conditions (cold case) results in a larger LFL and potential flash fire than accident case conditions. All other trends are similar for distance to LFL/flash fire across all test cases.

While flash fires are a function of the cloud concentration exceeding the LFL around the release, jet fires are a function of the discharge of flammable chemical from a pressurized system. Jet fires can occur with gas, liquid, or two-phase flow of flammable chemicals. Jet fires can only occur with the 1" and 3" release scenarios in this simulation due to the catastrophic scenario released instantaneously. The distances to the thermal radiation of 12.5 kW/m^2 are shown below for the flame lengths for each flammable liquefied gas, leak scenario, and test case.

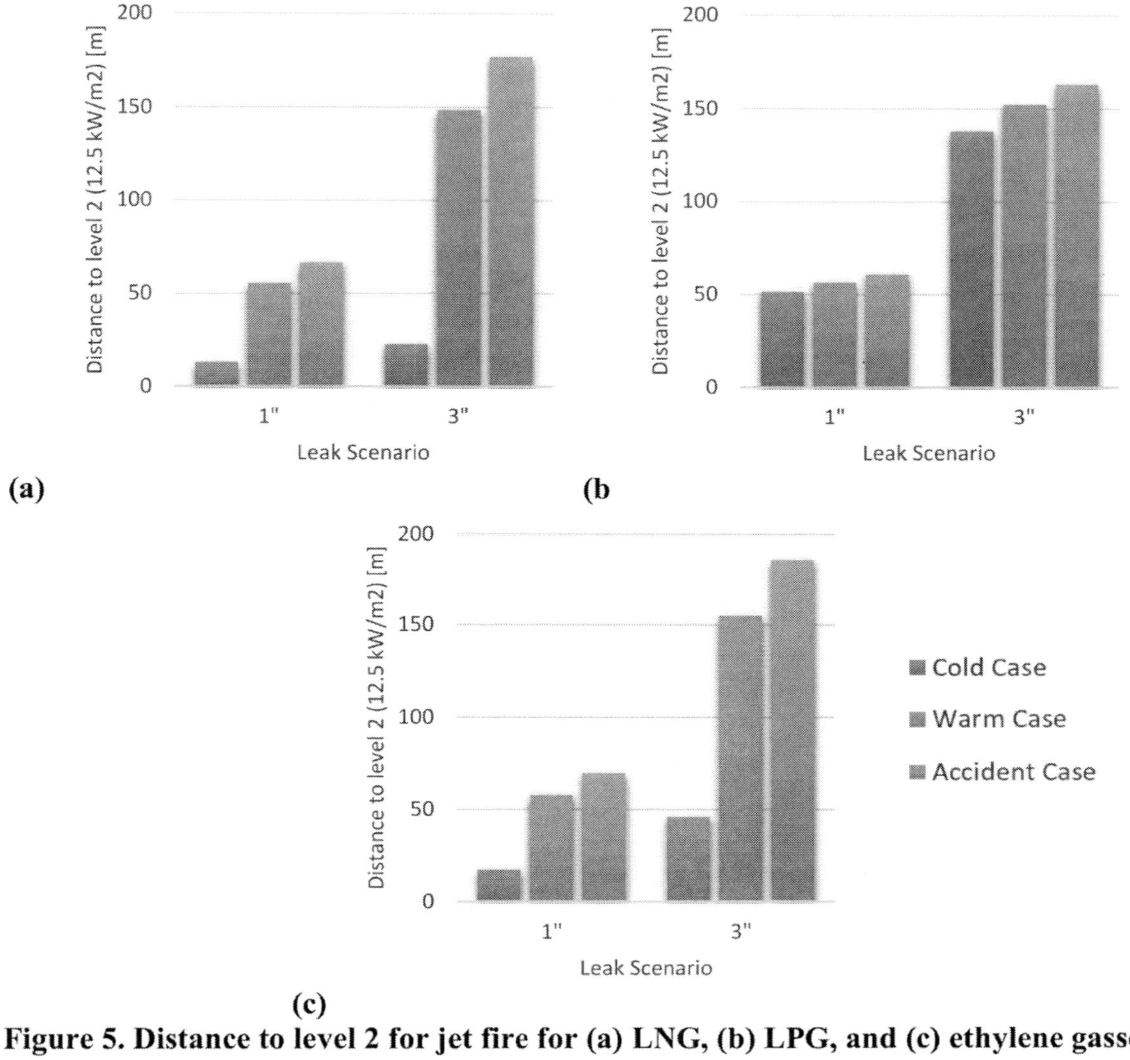

Figure 5. Distance to level 2 for jet fire for (a) LNG, (b) LPG, and (c) ethylene gasses for 1" and 3" leak scenarios for all temperature and pressure cases

The 3" leak provided larger flame lengths and thermal radiation distances than the 1" leak scenario across all flammable liquefied gases within the test cases. Variations in temperature and pressure shown in the different cases demonstrated a large effect on flame length and radiation distance. The largest risk across all chemicals and leak scenarios occurred under accident case conditions. Warm case conditions resulted in the next largest risk. Cold case conditions resulted in the smallest risk from smallest flame lengths and distances.

5 Summary

Typical consequence risk analysis calculation methods use the temperature and pressure of flammable liquefied gases under steady-state processes. In the case of an accidental release, such as mechanical impact after transit, the temperature and pressure are not reflective of steady-state conditions. Three flammable liquefied gases, commonly transported across rail and road, were evaluated for three leak scenarios and investigated under three realistic temperature and pressure combinations: cold—loading conditions, warm—lengthy transport conditions, and accident—impact after loading conditions. The results of this study reveal that temperature and pressure impact the potential consequences and risks of flammable liquefied gases released with respect to discharge dynamics, pool fires, flash fires, and jet fires. For an accurate consequence and risk analysis, it is vital to take into consideration the temperature and pressure conditions at those non-steady state conditions that occur in loss of containment.

6 References

[1] Pope M, Drewes J, May J. Generic hazard list for railway systems. In 7th World Congress on Railway Research, Montreal 2006.

[2] Tadros WA. Chair, Transportation Safety Board of Canada, Lac-Mégantic, Quebec, Release of Railway Investigation Report R13D0054, August 19, 2014.

[3] CCPS. *Guidelines for Chemical Process Quantitative Risk Analysis*. Center for Chemical Process Safety, American Institute of Chemical Engineers, New York, NY, 1999.

[4] Hart R, Morrison D, Garcia M. What is the Safest Way to Move LNG?

[5] Taylor Wharton Instructions Manual for LNG ISO Container, Available at http://twcryo.com/wp-content/uploads/2018/08/MANUAL-LNGT-40-FOOTER-VER.-1.pdf Accessed on January 16, 2019.

[6] Corban Energy Group LNG ISO Containers, Available at http://www.corbanenergygroup.com/lng-iso-containers/ Accessed on January 18, 2019.

[7] NFPA 59A: Standard for the Production, Storage, and Handling of Liquified Natural Gas (LNG), 2001. In *NFPA National Fire Codes Online*. Retrieved from http://codesonline.nfpa.org

[8] 49 CFR § 19.2059 (2015).

[9] Perry, Robert H., and Don W. Green. *Perry's Chemical Engineers' Handbook*. New York: McGraw-Hill, 2008.

[10] Kumar A, Saluja R, Pakala P. Development of a Numerical Model for Studying Flash Fire dynamics. Toledo, OH. 2001.

19 AIChE Spring Meeting
15TH GLOBAL CONGRESS ON PROCESS SAFETY
NEW ORLEANS • 2019

CFD-Based Probabilistic Hazard Analysis as an Early Tool to Improve FLNG Design

Scott Davis
Gexcon US
4833 Rugby Avenue Suite 100
Bethesda, MD
sgdavis@gexcon.com

Drew Botwinick
Gexcon US
Houston, TX

Prepared for Presentation at
American Institute of Chemical Engineers
2019 Spring Meeting and 15th Global Congress on Process Safety
New Orleans, LA
March 31 – April 3, 2019

CFD-Based Probabilistic Explosion Hazard Analysis as an Early Tool to Improve FLNG Design

Scott Davis
Gexcon US
4833 Rugby Avenue Suite 100
Bethesda, MD
sgdavis@gexcon.com

Drew Botwinick
Gexcon US
Houston, TX

Keywords: concept, pre-FEED, explosion risk analysis, consequence modeling, ventilation analysis, FPSO, FLNG

Abstract

Process areas on floating LNG facilities (FLNGs) are typically designed to be very compact, due to the high costs of "floating real estate". Given the high congestion levels of FLNGs, explosion design accidental loads (DALs) can become very large and are very sensitive to the layout of piping and equipment, the size of safety gaps, specifics of the liquefaction process, etc. For these reasons, the analysis of explosion hazards on FLNGs is a critical step that should accompany the facility development from the very earliest design phase. This paper will discuss how a probabilistic explosion hazard analysis methodology based on computational fluid dynamics (CFD) modeling can be used to optimize the layout of an FLNG facility with respect to DALs. The methodology uses a detailed 3D model of the FLNG geometry (down to 1-2 inch pipe diameters) within which gas dispersion and vapor cloud explosion scenarios are simulated using a CFD tool; leak rate probability data and time-dependent ignition model are then used to calculate probability of exceedance curves and to determine the DALs. The paper will present results from studies which compared the effectiveness of a number of design variables, such as:

- Different liquefaction processes;
- Docking distance between FLNG and LNG carrier;
- Module layout.

The paper will demonstrate how CFD-based probabilistic methodologies can provide FLNG designers with critical safety-related design implications previously not available until the later stages of design. More specifically, the advanced methodology implemented in the present work, which incorporates advanced techniques to supplement an "incomplete" model of the FLNG geometry in concept or design phases before front-end engineering design (FEED) to levels anticipated in the "as-built" condition, provides a powerful tool to assess FLNG layouts/technologies in the earliest design phase. This paper will provide some case studies on how valuable insights were gained by applying a detailed 3D model in early, concept phase of the design.

1 Introduction

Cost effective design of FNLG layouts is difficult to achieve, given the "floating real estate". Designers of these units are significantly restricted given these high prices. Therefore, these units are constructed with equipment clusters. In some cases, these units may also be present next to natural barriers that may affect how natural ventilation is distributed along the facility. Therefore, affecting the mixing of air and fuel released and the severity of the explosion. This requires an in-depth consequence and risk analysis to be performed to ensure that the severity and likelihood of having an explosion are minimized. Therefore, the need of a model that accounts for inclusion of geometry and obstruction should be a high priority for designers.

The layout and design decisions made during early design phases have a significant impact on flammable and toxic hazards that are often only assessed once the design is completed or in its later stages of development. The amount of re-engineering (and therefore cost) of mitigating the effects of these decisions increases as the project is developed, and eventually in construction or operation. As a result, one of the most effective strategies for limiting flammable or toxic hazards is to assess the impacts of design choices on the risk of these hazards at an early stage.

Most of the widely used models in early design phase do not account for detailed congestion and obstruction. As it has been shown by Dharmavaram [1], obstruction plays a key role in the overall dispersion of the cloud. For instance, given the source term conditions of a release, the dispersion of a vapor cloud in an open environment tends to reach longer distances compared to the dispersion of the same release in a congested environment. Therefore, the use of models that do not account for obstruction in the path of a vapor cloud may overestimate the overall size of the cloud. This will ultimately have an impact on the explosion scenarios that need to be studied further and in the overall safer and optimal design of the facility.

2 Anticipated Congestion

One major drawback during early design phase (e.g., concept or pre-FEED) is that the geometry is lacking sufficient detail to determine accurate dispersion and explosion consequences. This is especially true when using 3D CFD software for evaluating design accidental loads (DAL) on the various safety critical systems, where the explosion loads could be significantly lower if the necessary details in the geometry are not considered in the analysis. Given the costs associated with topside space, congestion levels and layout of modern FLNGs has become an increasing critical factor in managing hazards on such vessels. Figure 1 shows an example of elevated congestion levels on the most recently built FLNG, Shell's Prelude.

Figure 1: Shell's Prelude FLNG to provide perspective on realistic congestion-
(photo source: https://www.shell.com/about-us/major-projects/prelude-flng.html)

The benefit of more accurate modeling comes with the drawback that a 3D model is required—and 3D CAD models of facilities are typically incomplete during the early stages of design, as illustrated in Figure 2. When congestion due to small components is not integrated into the explosion pressure analysis model, the design blast loads will be severely underestimated. Previous studies concluded that explosion pressures can differ by an order of magnitude between an early-phase CAD model and the final, detailed design [2].

The power of being able to accurately evaluate facilities in early design phase is that the layout and the distribution of equipment can be easily modified as nothing is built yet. Therefore, several possible layouts can be "tested" and optimized by minimizing the explosion risk. This can be performed by the evaluation of the effectiveness of the safety gaps, type of flooring used and the equipment distribution among others. The overall evaluation of these variables will help on minimizing the size of the cloud and severity of explosions; thus, making the FLNG facility safer. On the contrary, if the facility is already built, most of these variables will not be easily modifiable. For instance, it will be costly to relocate equipment as this means adding piping/cabling and safety gaps are already in place.

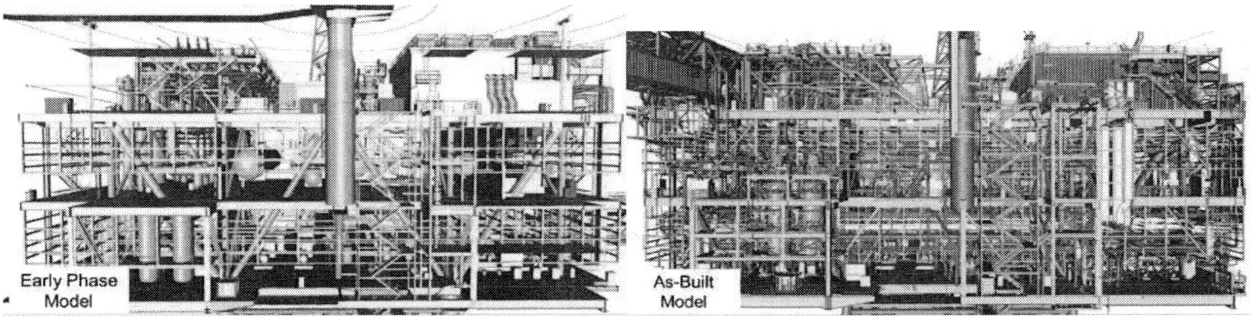

Figure 2: Illustration of difference in congestion levels between CAD models in the early design phase and in the final, as-built design [1]

Based on the information mentioned in previous paragraphs, one of the most effective strategies for limiting explosion overpressure hazards is to assess the impacts of design choices on explosion risk at an early stage. This requires methods that will provide accurate overpressure predictions for a completed installation well before the installation is completed. Determining explosion overpressure risk is complex. The risk is affected by ventilation conditions, flammable vapor cloud formation, and geometry/layout configuration, in addition to the more directly related cloud position and ignition location.

As explosion consequences are extremely geometry dependent and to provide meaningful explosion loads in early design phase, Gexcon has developed and employs a methodology called the Anticipated Congestion Method (ACM). This methodology involves a fairly detailed assessment of a wide range of project documentation (PFD, P&IDs, Equipment lists, etc.) as well as comparison with model characteristics from similar installations. To this end, extensive research has been performed over the last 20 years into methods on how to supplement early-phase CAD models with anticipated congestion to achieve congestion levels of as-built facilities. This methodology has recently been verified and proven to yield reasonably accurate congestion levels in order to successfully evaluate and optimize concept-phase designs [3][4].

Examples of an early design phase process module are provided in the left images of Figure 3 and Figure 4, while the right images include the structural and process components anticipated to exist in final as-built design (i.e., structural support, railing, relief piping, blowdown piping, instrumentation, water deluge, access platforms, etc.).

Therefore, prior to performing 3D probabilistic studies that can provide critical DALs, the concept or pre-FEED models must be supplemented with anticipated congestion to bring the geometry detail (from an explosion consequence point of view) to be representative of an "as-built" design.

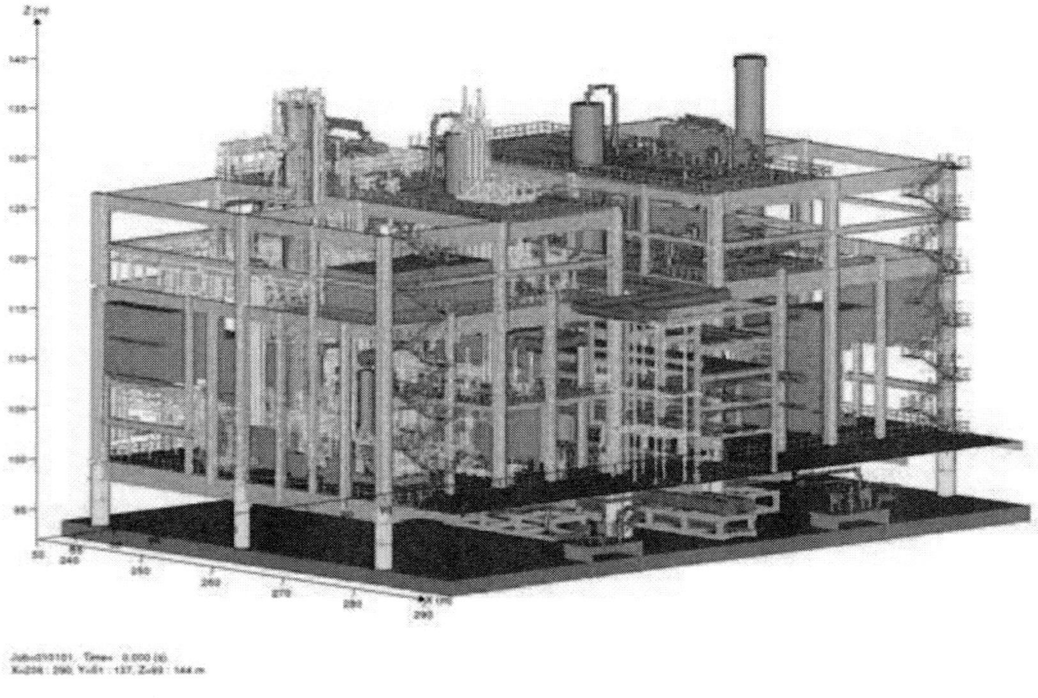

Figure 3: Example of an FLNG module both with ACM (left) and at concept design phase (right)

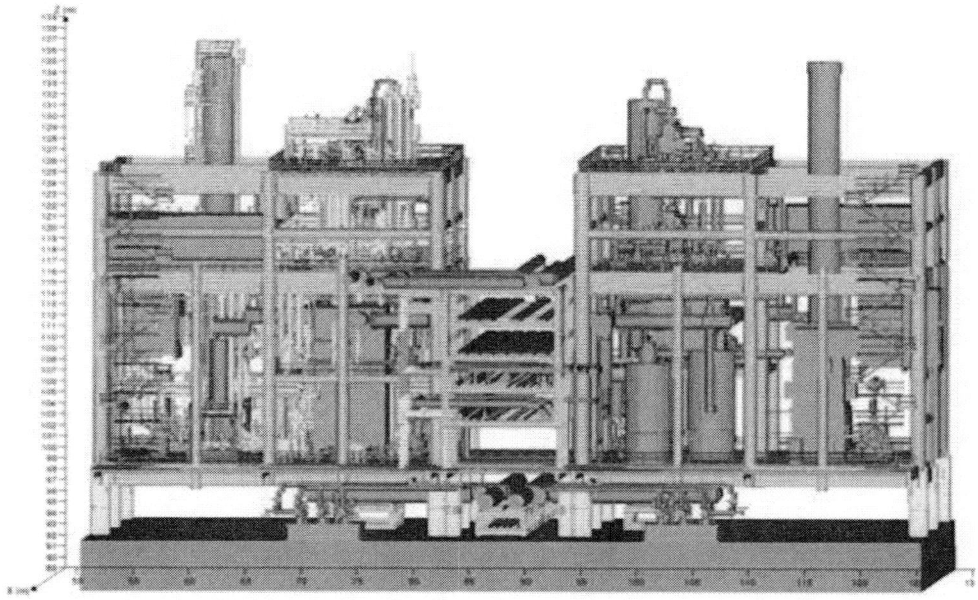

Figure 4: Example of an FLNG module both with ACM (left) and in concept design phase (right)

3 Different Liquefaction Technologies

Especially in the concept phase of a FLNG, it is reasonable to evaluate different liquefaction technologies. Selecting different technologies has implications on efficiency, safety, and operational costs. It is possible to evaluate the safety implications of selecting different liquefaction technologies using a risk-based approach, even in early design, by making use of methods such as ACM.

Figure 5 shows two views of a FLNG concept vessel—the two vessels are meant to be nearly identical except each is equipped with different liquefaction technologies. Implementing each of the two technologies requires different amounts of piping, equipment arrangement, and refrigerant compositions. The 3D models for each alternative were developed in order to compare the safety implications of each design.

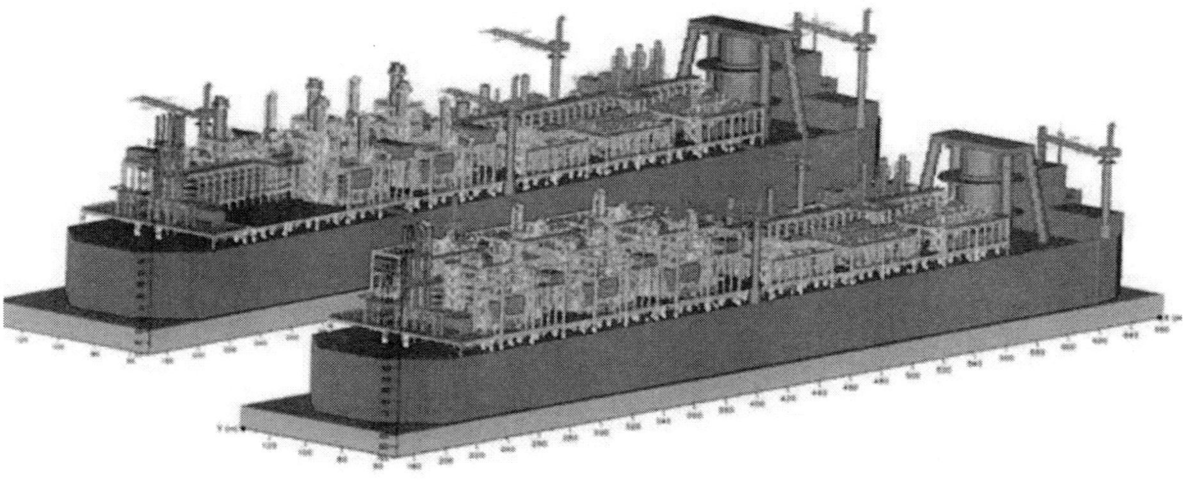

Figure 5: Side-by-side view of otherwise identical FLNG vessels using two different liquefaction technologies

When developing the risk-based approach, a critical step is to develop representative release frequencies for the different liquefaction technologies to properly account for leaks. In addition to developing the two 3D models, the design and piping changes were applied to establish representative leak frequencies for both designs. Figure 6 shows the accumulated leak frequencies for the two technologies split out by liquid and vapor releases. The leak frequencies associated with each liquefaction technology were ultimately fairly close.

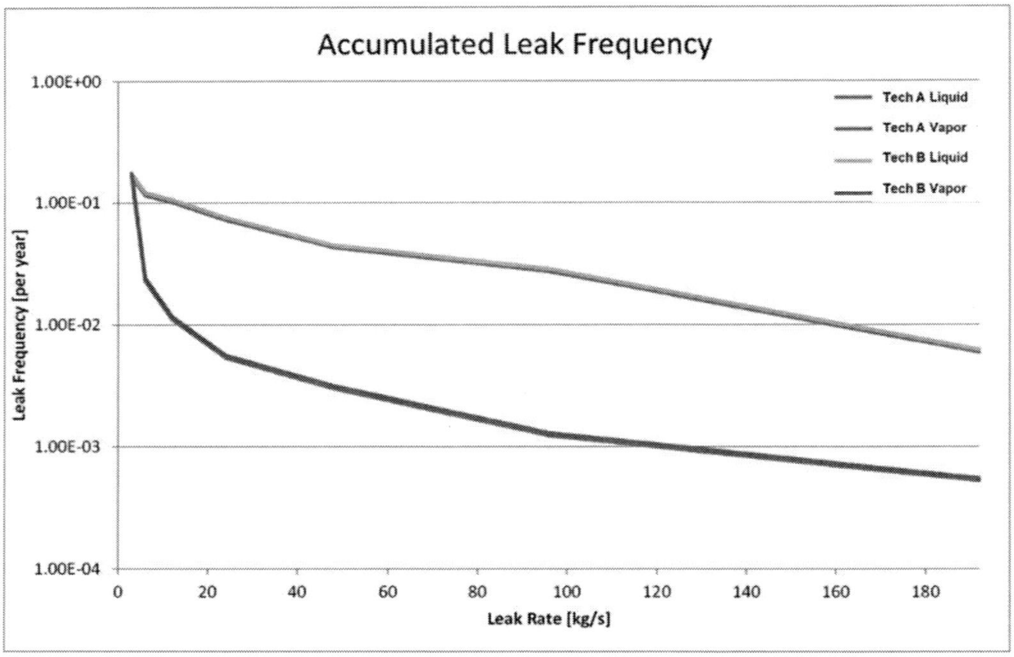

Figure 6: Accumulated leak frequencies for two different liquefaction technologies—each split into liquid and vapor releases

"Technology B" involves less congestion and confinement—a fact that shows in the ventilation results in Figure 7. The vessel designs call for using weather vaning, maintaining the vessel orientation into the wind, rotating around the turret. As such, all of the wind directions are relative to zero degrees—i.e. head on to the vessel. The lowest ventilation occurs for the wind direction of zero when all of the modules on the vessel are aligned, and flow resistance is the highest as the wind was blowing through each of modules along the vessel.

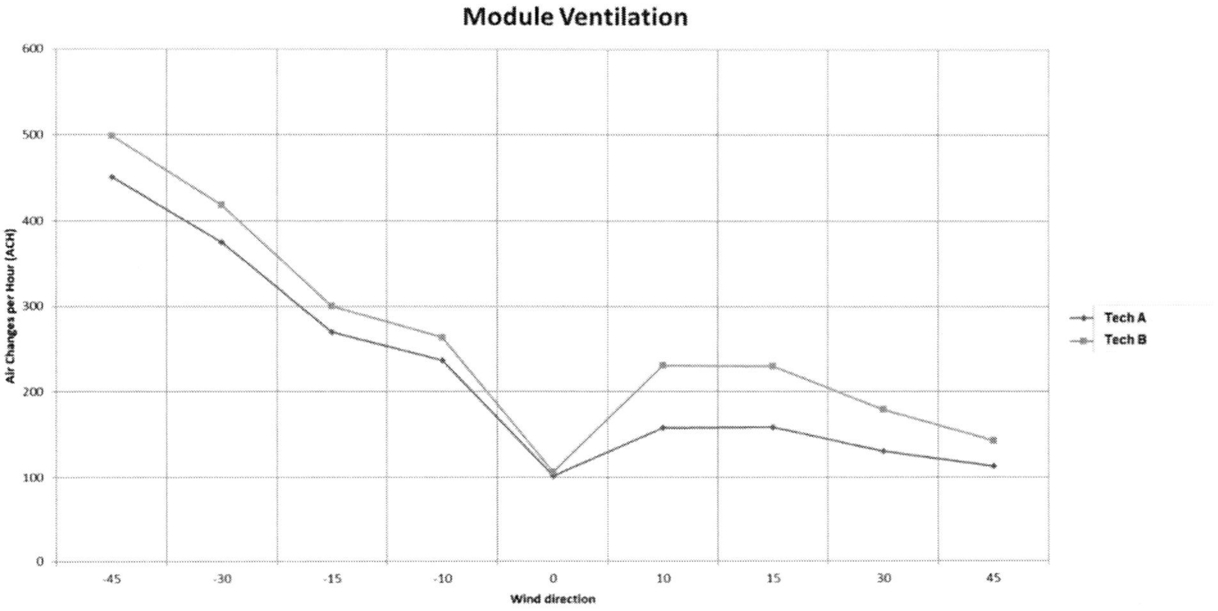

Figure 7: Ventilation results for the two different liquefaction technologies

Accounting for the ventilation conditions, a range of refrigerant release scenarios were modeled using CFD dispersion. The results of dispersion scenarios are presented in Figure 8 and Figure 9. It is noteworthy that "technology B" produced large clouds, especially the average cloud sizes in the smaller leak rates (3 to 12 kg/s) which would be associated with smaller hole sizes and therefore expected to be more frequent. However, "technology B" is also the less confined and congested design, as demonstrated by the ventilation results. The results for "technology A" also show comparisons of two different minor layout alternatives (use of plating versus grating).

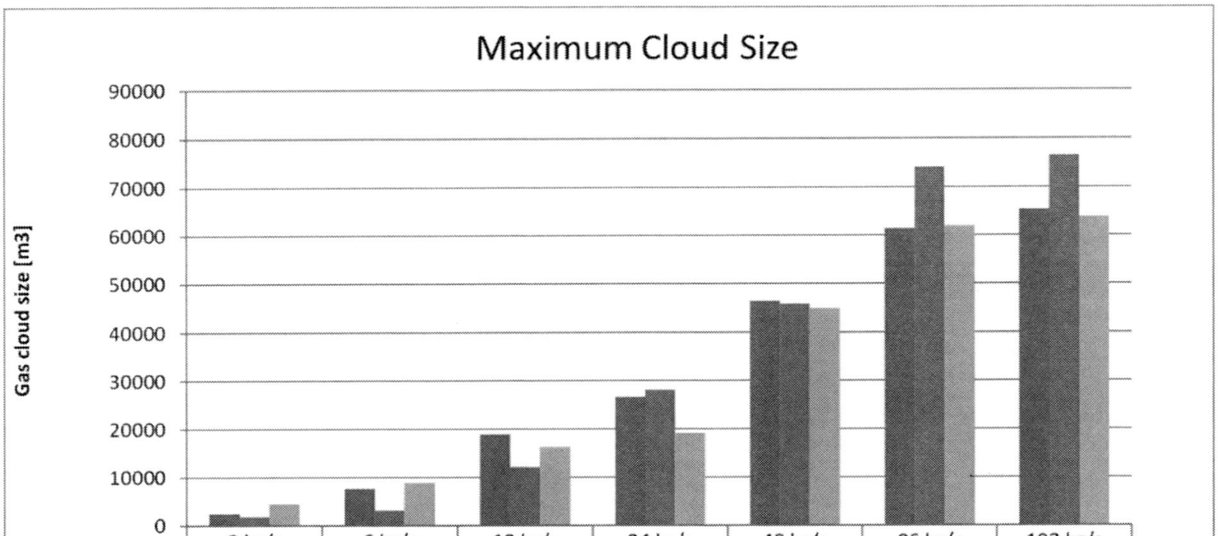

	3 kg/s	6 kg/s	12 kg/s	24 kg/s	48 kg/s	96 kg/s	192 kg/s
Tech A (Plated)	2502	7680	18945	26646	46379	61420	65367
Tech A (Grated)	1836	3142	12143	28094	45795	74089	76453
Tech B	4498	8868	16282	19143	44939	61923	63842

Figure 8: Maximum cloud sizes for two liquefaction technologies from dispersion simulations

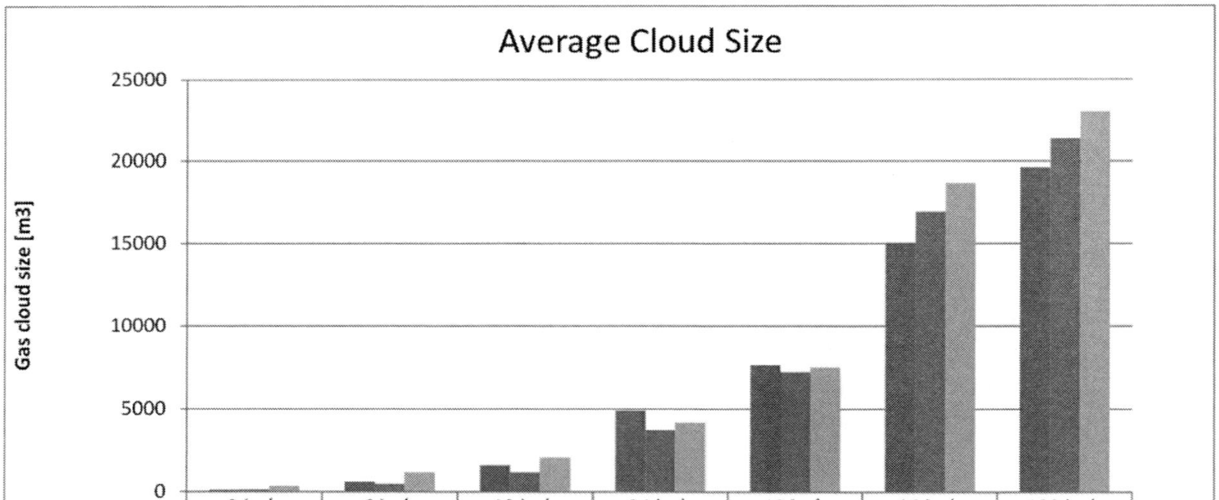

	3 kg/s	6 kg/s	12 kg/s	24 kg/s	48 kg/s	96 kg/s	192 kg/s
Tech A (Plated)	109	582	1579	4906	7668	15032	19607
Tech A (Grated)	125	464	1149	3733	7250	16927	21346
Tech B	342	1159	2052	4178	7528	18690	22953

Figure 9: Average cloud sizes for two liquefaction technologies from dispersion simulations

A range of CFD explosion simulations were also run to establish the relationship between overpressure and cloud size. The results are illustrated in the plot in Figure 10.

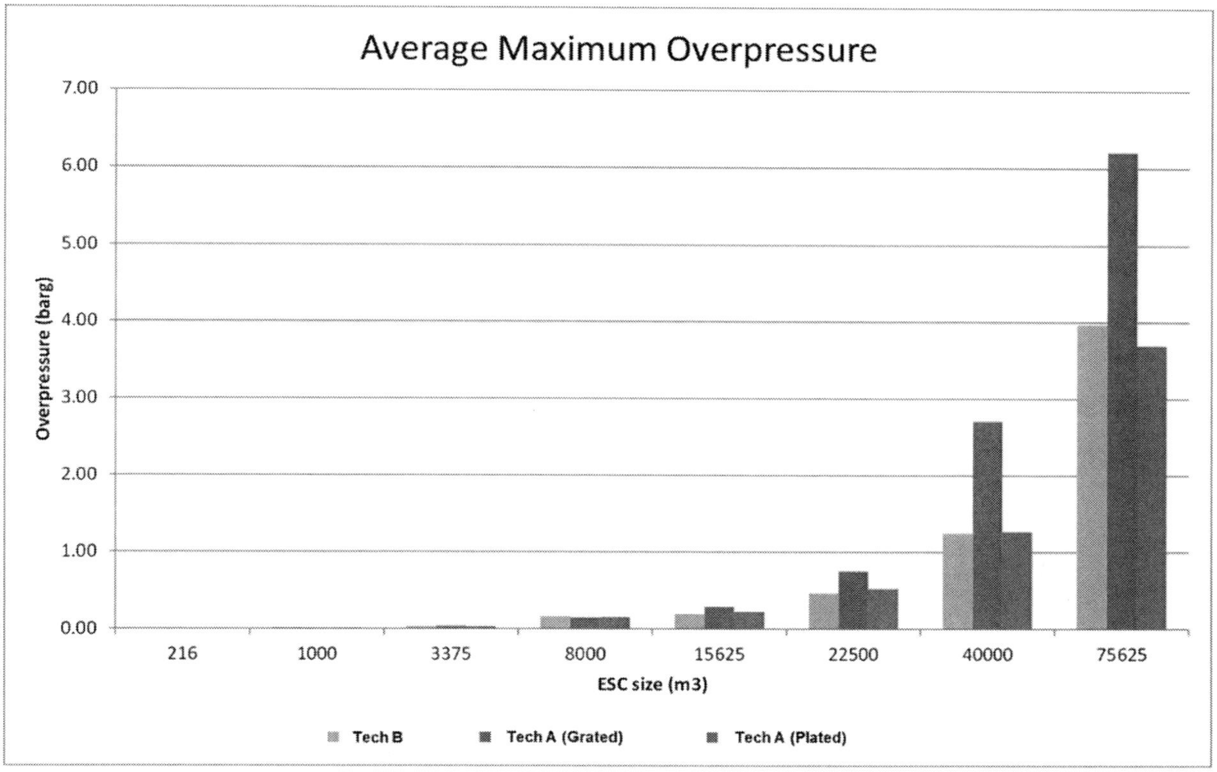

Figure 10: Average maximum overpressure of two different liquefaction technologies

After discounting the grated variation of "technology A" due to the high pressures, the results are generally similar for all cloud sizes when comparing the technologies. However, for the largest cloud size, the overpressure from "technology B" simulations are about 10% higher.

Ultimately, "technology B" wins out from an explosion risk perspective. "technology B" resulted in an overpressure on the main deck of 0.3 barg at a frequency of 10^{-4} per year, while "technology A" (plated) resulted in an overpressure of 0.8 barg at the same frequency on the main deck. Although "technology B" can lead to larger clouds for smaller, more frequent leaks, that is offset by the decrease in the confinement and congestion.

4 Separation Distance between LNG Carriers and FLNG

It is a common practice to have an LNG carrier pull alongside a FLNG for (off)loading. One of the safety concerns is the safe separation distance between the LNG carrier and FLNG. Shorter distances involve simpler logistics and lower costs at the trade-off that the LNG carrier needs to remain closer to the FLNG for the duration of the docking and offloading procedure. Figure 11 illustrates two cases for offloading to an LNG carrier. The 5m case is the simpler design case while the 15m spacing case requires more complicated docking and offloading procedures and potentially the construction of an additional docking platform between the FLNG and carrier.

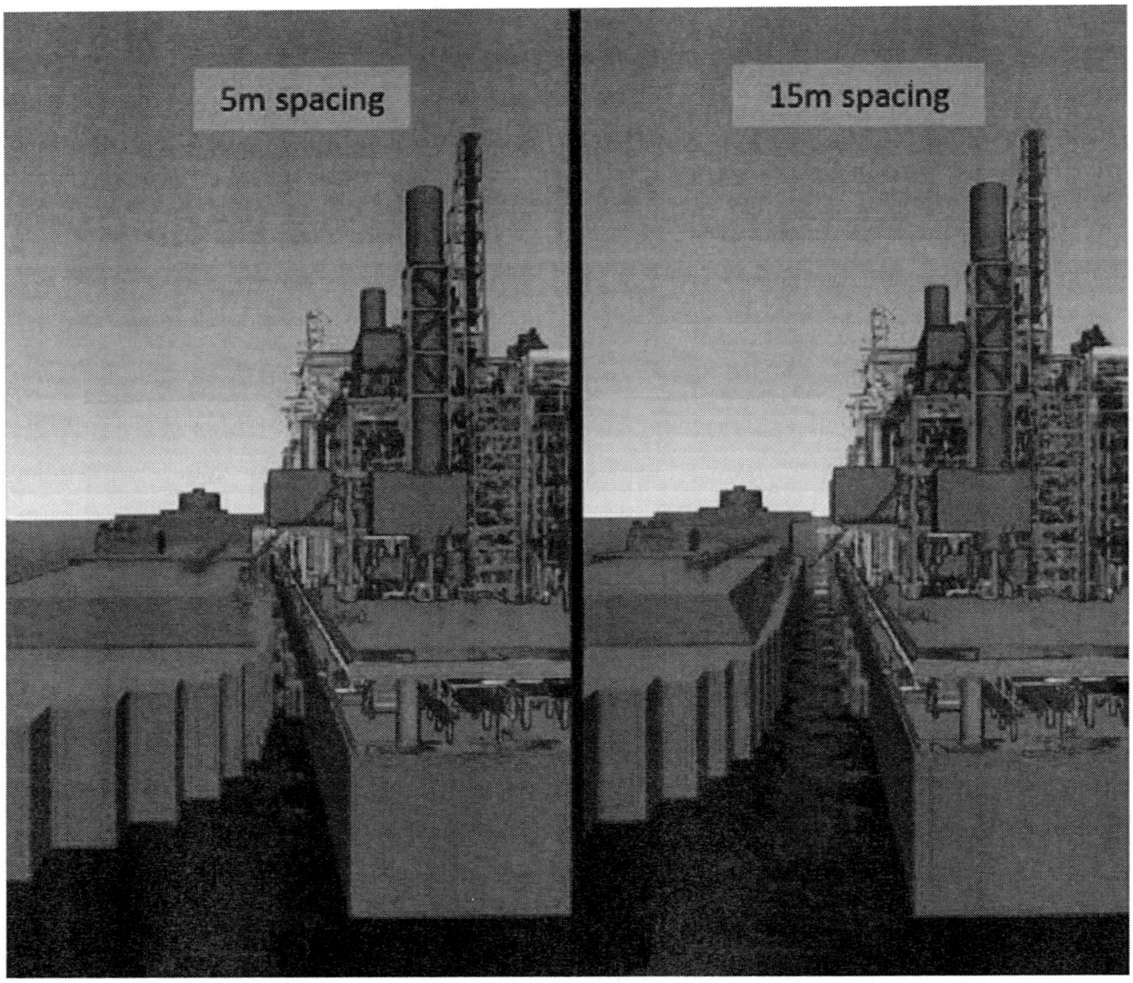

Figure 11: LNG carrier alongside a FLNG for two different distances between them (5m and 15m)

CFD simulations of dispersion and explosion events can be used to make decisions about the distance that would need to be maintained between the FLNG and LNG carrier. This is critical safety information that also will have a significant operating cost impact. Figure 12 shows flammable clouds formed by refrigerant spills from the FLNG liquefaction modules to the LNG carrier.

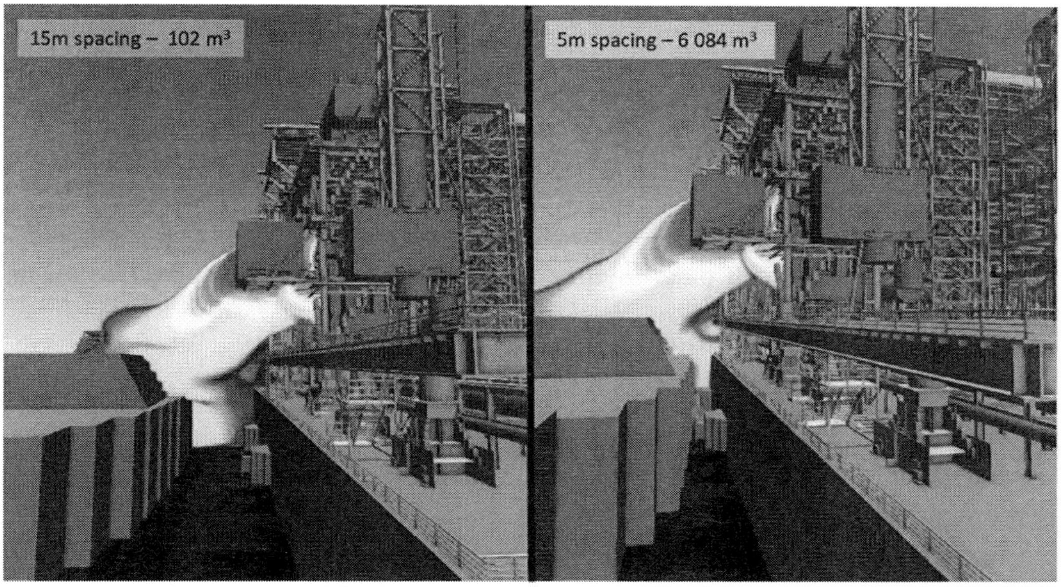

Figure 12: Refrigerant releases from liquefaction module to LNG carrier at 5m and 15m spacing

Another concern is the channeling effect between the LNG carrier and FLNG. For example, in slow wind conditions, there is a chance that the channel created between the carrier and FLNG will facilitate gas from the rapid phase transition accompanying an LNG spill onto water would be carried into the inter-deck or tank-top space, which is typically very congested and confined. For this example, with slow wind speeds, a significant gas cloud can build up in the tank-top space for the 5 m spacing presenting a significant explosion hazard for both the FLNG and carrier. This is illustrated in Figure 13.

Figure 13: LNG spill for 5m and 15m spacing

Explosion simulations were focused on explosions originating from the inter-deck space that could affect the LNG carrier. Explosion simulations showed the effect of the increased confinement caused by the proximity of the LNG carrier. Depending on the appropriate representative scenarios and risk acceptance criteria, it may not be possible to dock the carrier only 5 m away from the FLNG.

Figure 14 shows the pressures measured at the LNG carrier for a 3336 m^3 cloud located near the edge of the tank-top space. This type of cloud is among the possible scenarios that could occur from the release of cold, dense refrigerants from the topsides process during slower wind conditions. As shown in the figure, the pressures were reasonably high near the waterline of the LNG carrier for the 5m spacing case.

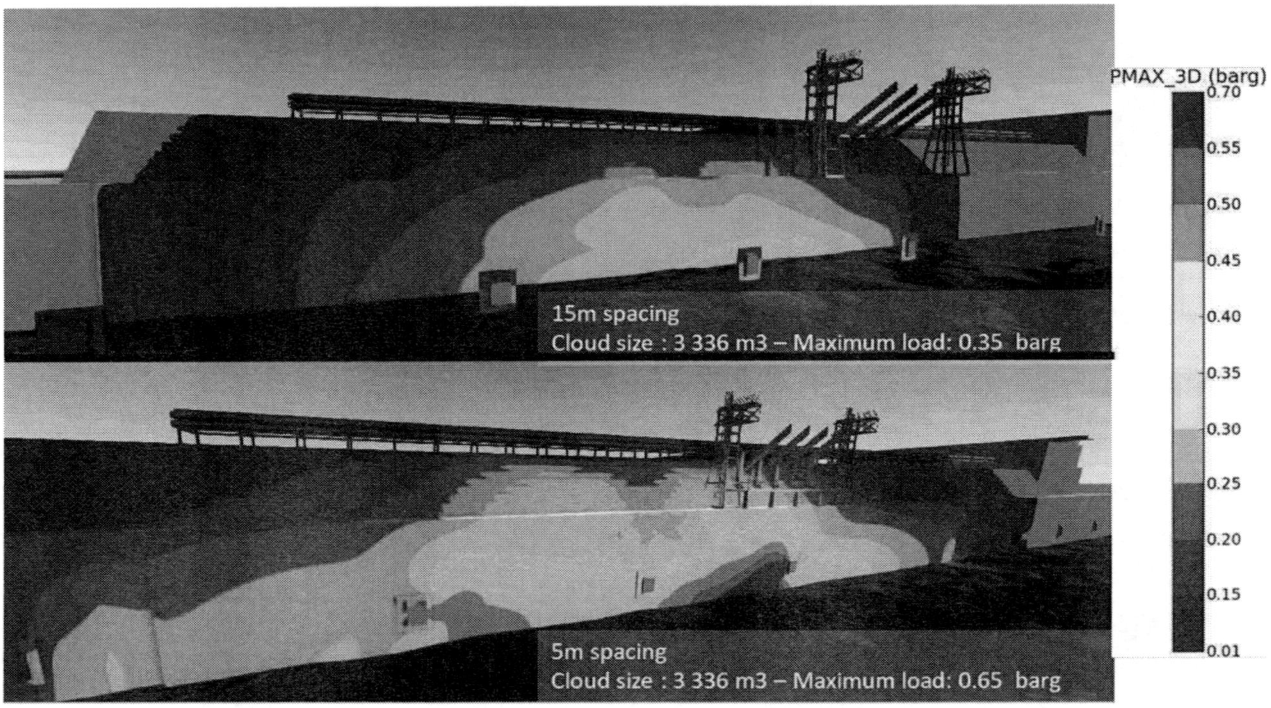

Figure 14: Explosion simulations for 5m and 15m spacing between the FLNG and LNG carrier

Using a risk-based approach and further CFD simulations, it is possible to optimize the distance between the LNG carrier and FLNG to meet risk acceptance criteria while minimizing the distance.

5 Module Layout

CFD ventilation studies can also be used as part of selecting and/or optimizing layouts. Ventilation studies are often performed to validate that layouts meet industry criteria or as a step in more complex safety studies; however, their value in making critical decision decisions in early-design that can drastically alter the risk posed by toxic, fire, and explosion hazards is often overlooked.

Many ventilation studies provide results as a single measure, such as "adequate ventilation" measured by the number of air changes per hour (ACH) for a given area. For example, certain hazardous area classifications require that the ventilation must be greater than 12 ACH at least 95% percent of the time (95th percentile ACH). However, the 95th percentile ACH only describes a limited range of the ventilation conditions at the facility and should not be used as the sole means of comparing alternative layouts or making design decisions.

A carefully constructed ventilation study that accounts for the nature of facility hazards can provide relatively fast and inexpensive insight into potential safety-oriented optimizations during early stages of facility design. This insight in the early design phase allows for facilities to be rigorously designed for safety—minimizing costly late-stage design changes. For example, it would not be practical to move around process modules after investing months in designing piping and cabling; however, changing the arrangement of process modules is more practical in early-design. In this example, we'll look at the effect of rotating a concept cylindrical FLNG in order to maximize ventilation in a process area.

Figure 15 depicts a conceptual, cylindrical FLNG design. Cylindrical designs have been used in various oil and gas applications; however, it is newer to the FLNG space. CFD-based ventilation studies enable reliable answers for new concepts.

Figure 15: Cylindrical FLNG approved-in-principle by ABS in 2014 (image: Sevan Marine)

Although the vessel is cylindrically shaped, it does not have radial symmetry, and therefore, the orientation of the vessel relative to the wind conditions will affect the ventilation patterns on the vessel. Figure 16 shows the ACH exceedance curves varying the orientation of the vessel.

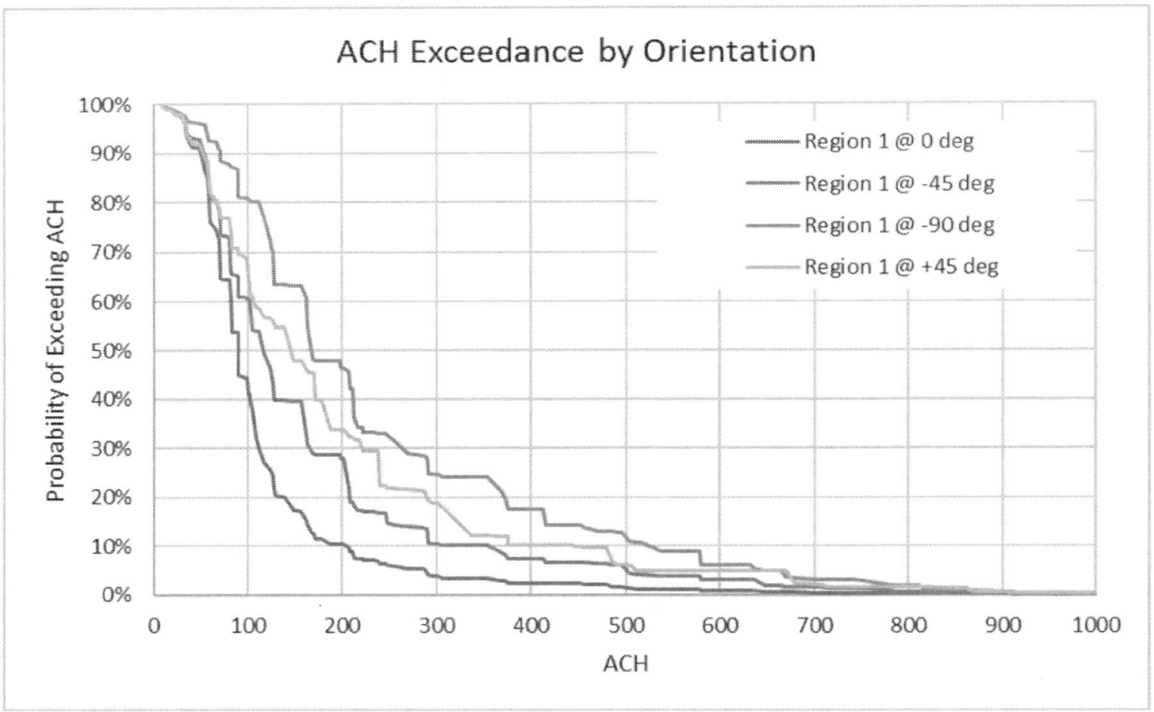

Figure 16: ACH exceedance curve for a range of different vessel orientations

The plot shows that rotating the vessel has a significant effect on the ventilation patterns in the region, measured as air changes per hour. Table 1 shows the 95th percentile ACH for the different vessel orientation. Matching with the exceedance curve plot, there is clearly one orientation that has significantly better ventilation than the others.

Table 1: 95th percentile ACH for different vessel orientations

ACH Curve for Region 1 Rotation from north	95th Percentile ACH
0° ↑	35
-45° ↖	35
-90° ←	57
+45° ↗	35

For this example, the 95th percentile result and the overall result, based on evaluating the area under the exceedance curve, are aligned. However, that is not always the case. Solely relying on the 95th percentile ACH can sometimes be misleading. Reviewing the values in the table shows that three out of four of the results are 35 ACH.

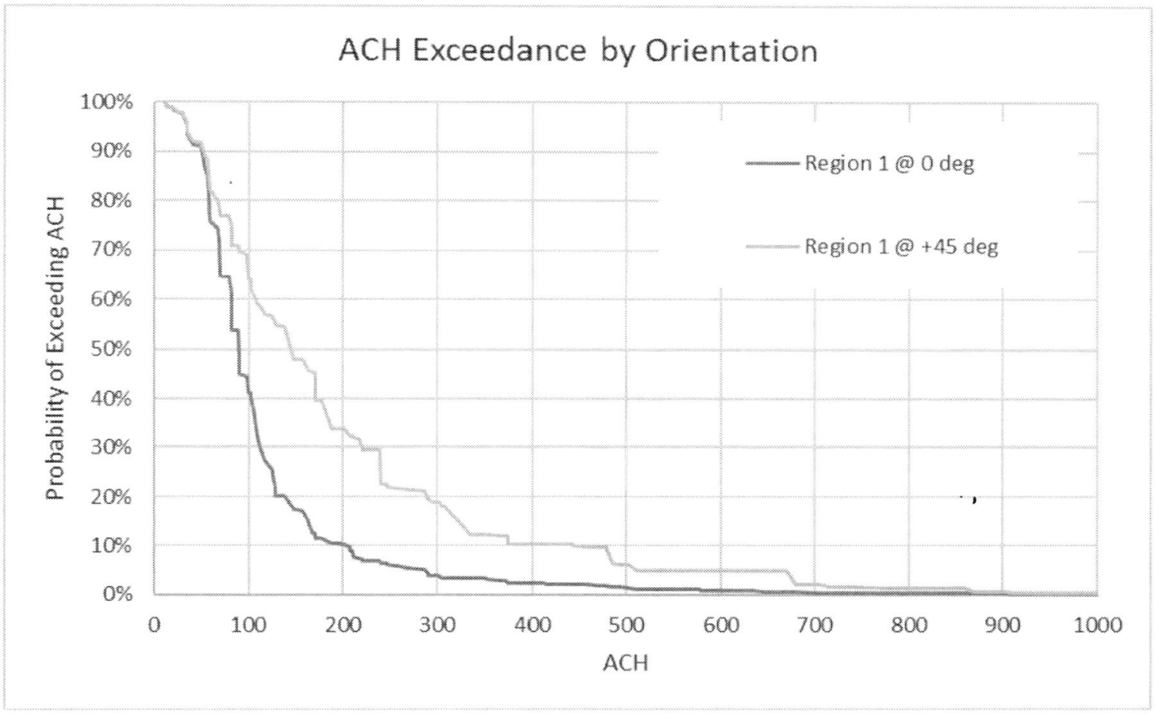

Figure 17: ACH exceedance curve for a subset of the previously displayed vessel orientations

Figure 17 shows two of the orientations that result in 35 ACH. Although the 95th percentile ACH is similar between these two orientations, there is a significant difference in the overall ventilation pattern. The approach taken to evaluate the ventilation patterns for rotating the vessel can also apply to individual process modules or equipment within process modules.

6 Stagnation Zones

In order to get better insight into the details inside a congested or process region, it is typical to provide a few velocity vector plots of the prevailing wind conditions, the worst wind conditions, or other conditions depending on how they will be used. These velocity plots can be used to gain insight into areas of stagnation or if there are particular flow paths that may be favored. This information can be indispensable for early consideration of muster locations and escape routes. (Although it is important to keep in mind that this information doesn't replace proper studies to verify these decisions.) Figure 18 shows a sample velocity vector plot.

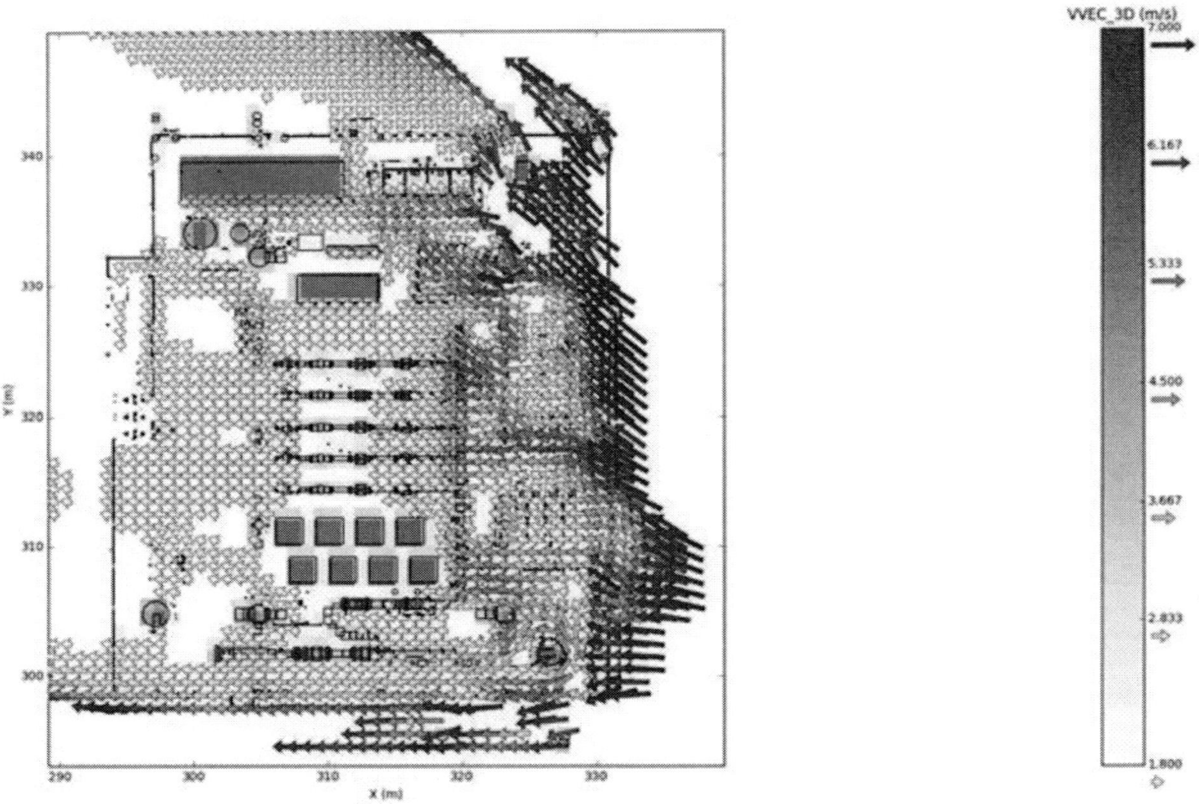

Figure 18: Sample velocity vectors plot in a cut-plane.

Although velocity vector plots can provide great insight into the flow conditions within regions of interest, it makes sense to leverage the statistical exercise used to generate the ACH exceedance curves to provide more useful insights. As discussed in the previous section, there is significantly more value to gain from ventilation studies than just a single measure of the 95th percentile ACH.

The entire set of CFD simulations can be combined with the wind probability data to provide composite views to identify stagnation zones in various regions of the facility. These plots can provide significantly better insight into early stage decisions regarding muster locations, escape routes, etc. Figure 19 and Figure 20 are representative examples of such depictions.

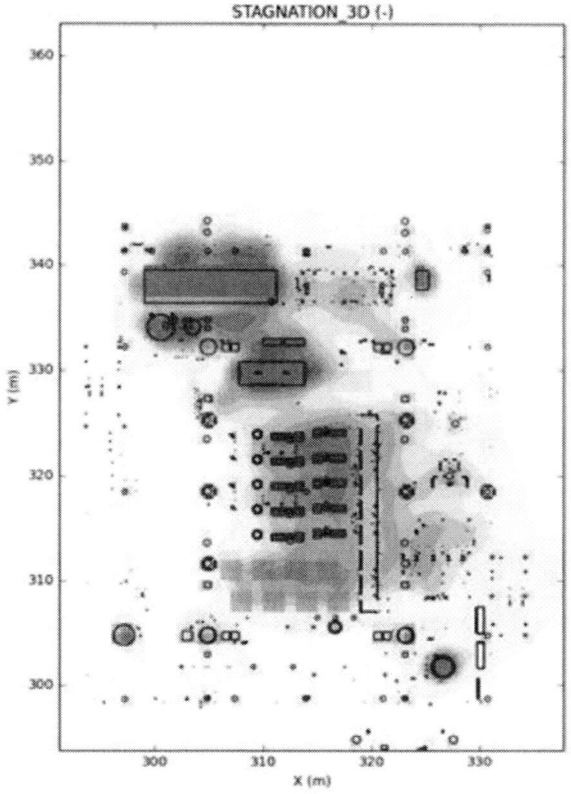

Figure 19: Sample 2D composite stagnation plot

Figure 20: Sample 3D composite stagnation plot

7 Conclusion

Given the level of congestion and constrain of space available for a FLNG facility, advanced tools to perform consequence analysis are recommended. These tools should be capable of integrating geometry obstruction and confinement, as these variables play a key role in the dispersion, *e.g.*, development of the cloud, and explosion. It is worth mentioning that the design phase is a pivotal point to minimize the explosion risk and maximize personnel safety. This could not be easily done once the facility is already built.

Industry capability to represent as-built facilities using early CAD models is reasonably well-developed. This enables using CFD methods to facilitate answering complex questions regarding ventilation, dispersion, fire, and explosion hazards with a reasonable degree of certainty, and therefore, providing useful insight into safety decisions early in the design process when these decisions can have the largest impact for the smallest cost.

8 References

[1] S. Dharmavaram, S. R. Hanna and, O. R. Hansen. "Consequence Analysis – Using a CFD Model For Industrial Sites" Process Safety Progress, Nov 9, 2005, DOI: 10.1002/prs.10068

[2] D. Botwinick and S. Davis, "Getting It Right for Projects in Early Design Phase: Advanced Techniques to Ensure Accurate Explosion Assessments," Process Safety Progress, Dec 5, 2018, DOI: 10.1002/prs.11995.

[3] Davis, S. "Development of Advanced CFD Tools for the Enhanced Prediction of Explosion Pressure Development and Deflagration Risk on Drilling and Production Facilities". RPSEA, NETL, DOE. 2017.

[4] Health & Safety Executive. Explosion Pressures Evaluation in Early Project Phase. HSE. 1999.

Time to Consequence Simulations in LNG Facility

Haribabu Chittibabu
Senior Process Engineer
Bechtel Oil, Gas & Chemicals, Inc.
3000 Post Oak Blvd., Houston, TX
hchittib@bechtel.com

Sameera Thakore
Senior Process Engineer
Bechtel Oil, Gas & Chemicals, Inc.
3000 Post Oak Blvd., Houston, TX
ssthakor@bechtel.com

Jegan Babu Arumugakan Thuraiswamy
Deputy Chief-Process Safety
Bechtel Oil, Gas & Chemicals, Inc.
3000 Post Oak Blvd., Houston, TX
jbarumug@Bechtel.com

Prepared for Presentation at
American Institute of Chemical Engineers
2019 Spring Meeting
April 01-04, 2019

UNPUBLISHED

Time to Consequence Simulations in LNG Facility

2019 AIChE Spring Meeting.
Topical 6: 19th Topical Conference on Gas Utilization
LNG Safety

Key words

LNG (Liquefied Natural Gas), Safety Instrumented Function (SIF), Safety Integrity Level (SIL) Dynamic simulation, Time to Consequence (TTC), IEC (International Electrotechnical Commission)

1.0 Abstract

Process safety has always been a key issue during the design and operation of a LNG facility. Safety Instrumented Function (SIF) with a specified Safety Integrity Level (SIL) is necessary to achieve functional safety (IEC Standard 61511).

Safety Instrumented Function is the action that a safety system takes to make the overall process safe during emergency and abnormal operations. Safety Instrumented Function may include only a portion of a Safety Instrumented System (SIS) or it may include multiple SIS's. Safety Instrumented System is a set of hardware and software that is designed to perform set of actions to confine the process within safe limits.

Time to Consequence (TTC) or Process Safety Time (PST) is the period of time, during a process deviation event, between the time that a failure occurring in the process or the basic process control system and the time at which the safety hazard is realized assuming that the Safety Instrumented System somehow fails to execute the final actions required by the Safety Instrumented Function.

> According to the CCPS *Guidelines for Safe and Reliable Instrumented Protective Systems* Process Safety Time (PST) is *"the time period between a failure occurring in the process or its control system and the occurrence of the hazardous event."*
>
> API 556 second edition, 2011: *"the interval between the initiating event leading to an unacceptable process deviation and the hazardous event"*

Per IEC 61511 US Version: ANSI ISA 84.00.01: Given the degree of uncertainty in the process safety time, the Safety Instrumented System should be capable of completing its action within one-half of its allocated process safety time. TTC $\geq$ 2 x IRT. IRT is the Instrument Response Time (Sensors, filters, scan time, Final Control Element).

A Fail Safe analysis was done for LNG Facility for all the interlocks to determine if the response of the Safety Instrumented System is quick and fast enough to protect the equipment or for containment. As a result of this, process engineers had to come up with Time to Consequence for different interlocks.

Certain Time to Consequence calculations require dynamic analysis of the system protected by Safety Instrumented Function. Dynamic simulation utilizing rigorous mathematical models has become an influential tool in the process design, design validation, control system verification, startup support and troubleshooting. Many recent developments in the simulation software and technology has led to the development of large scale dynamic models and their use in the entire project life.

In this paper, how an existing dynamic simulation of the integrated LNG facility was used to determine the process safety time or TTC of safety critical interlocks to support the SIS design, the steps involved, the methodology and how insufficient TTC has been addressed will be discussed.

2.0 Introduction and Definition

Safety Instrumented Function (SIF) with a specified safety integrity level is necessary to achieve functional safety. A safety instrumented function can be either a safety instrumented protection function or a safety instrumented control function. SIF is the action that a safety system takes to make the overall process safe when confronted with a potential hazard. A single SIF may include only a portion of an SIS or it may include multiple SIS's.

Safety Instrumented System (SIS) is set of hardware and software that is employed to perform a set of actions. Safety Integrity Level (SIL) is a discrete level for specifying the safety integrity requirements of the safety instrumented functions to be allocated to the SIS. SIL level 4 has the highest level of safety integrity; SIL level 1 has the lowest. (IEC Standard 61511)

SIL Levels: 0 1 2 3 4

Require TTC or PST Calculations (levels 1–4)

Time to Consequence (TTC) or Process Safety Time (PST) is the amount of time that is available to take action against the process disturbance (or failure) to move the process to a safe state after an out-of-control condition has occurred in the process. The key to this term is that it is completely PROCESS related, and the SIS used has no bearing on the value. TTC can vary greatly between processes.

Figure 1 shows a graphical illustration of TTC or PST.

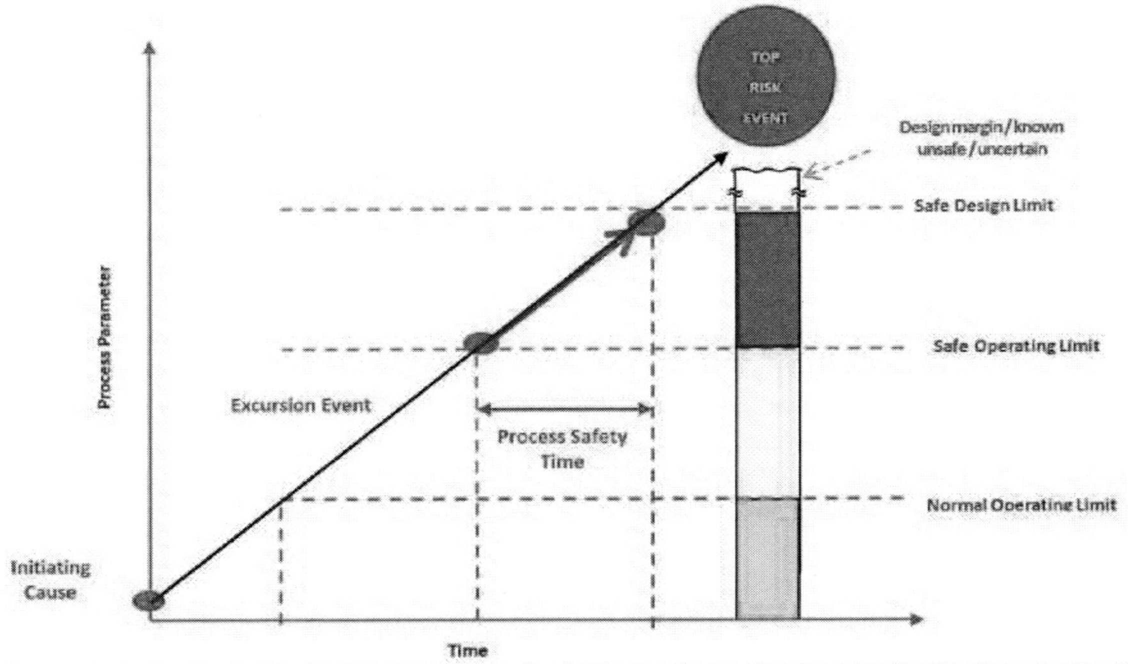

Figure 1: TTC or PST Illustration

Examples

TTC for Compressor: When the flow rate through a compressor becomes very low, material can back flow through the compressor causing it to 'surge'. In this case, the time to consequence is the interval between the process failure that leads to the low flow (alarm set point) through the compressor and when the compressor actually surges. For this example, the "time to consequence" is measured as a few milliseconds.

TTC for fired process heater. If the skin temperature of the heater tubes goes high, the tubes can rupture, causing the process fluid to be released into the heater firebox, potentially causing fire and explosion. In this example, the "time to consequence" is the time interval between when tube skin temperature reaches the higher limit of normal operating range due to the failure of process fluid through the tube or high burner fuel supply), and the point in time when the tube ruptures. In the worst case, this time will be minutes, and in the best case, this time interval could be weeks or months.

For design purposes, TTC is defined as the time interval between the initiator of an interlock (trip setpoint) identifying an out of control condition and the occurrence of the unwanted event. Time to consequence calculation always starts with trip setpoint (which initiates the interlock) and not the alarm setpoint, which would be less conservative and make TTC artificially larger.

TTC is important because it determines the maximum amount of time that can be allowed for the SIS to perform ALL of the following actions. Depending on the hardware that is used to perform the shutdown, more or less actions than described in the list below will have to be performed.

1. Sense the out-of-control condition
2. Digital filtering of input signal
3. Input process scan time
4. PLC program scan time
5. Any trip delay timers set to remove process noise must time out
6. Output process scan time
7. Digital filtering of output signal
8. Fully actuate the output device (some valves can require up to 30 seconds to fully stroke!)

3.0 Industry Standards

Important influential institutes within each industry create standards to improve safety across the industry they operate in. The International Electrotechnical Commission (IEC) is the primary institute that does this for all electrical, electronic and related technologies, also known as "electro technology". The American Petroleum Institute (API) is the institute that does this for the oil and natural gas industry. An important term is "Time to Consequence" (TTC), which in short means the time available between a moment in time that a failure occurs and the actual hazardous event caused by the same failure. Both of these institutes issued standards regarding SIF and determination of TTC.

According to the CCPS *Guidelines for Safe and Reliable Instrumented Protective Systems* Process Safety Time is *"the time period between a failure occurring in the process or its control system and the occurrence of the hazardous event."*
API 556 second edition, 2011: *"the interval between the initiating event leading to an unacceptable process deviation and the hazardous event"*

Per IEC 61511 US Version: ANSI ISA 84.00.01: Given the degree of uncertainty in the process safety time, the Safety Instrumented System should be capable of completing its action within one-half of its allocated process safety time. PST or TTC $\geq$ 2 x IRT. IRT is the Instrument Response Time (Sensors, filters, scan time, Final Control Element).

4.0 Time to Consequence Determination Steps

The following flowchart shows the information required to start the time to consequence calculations:

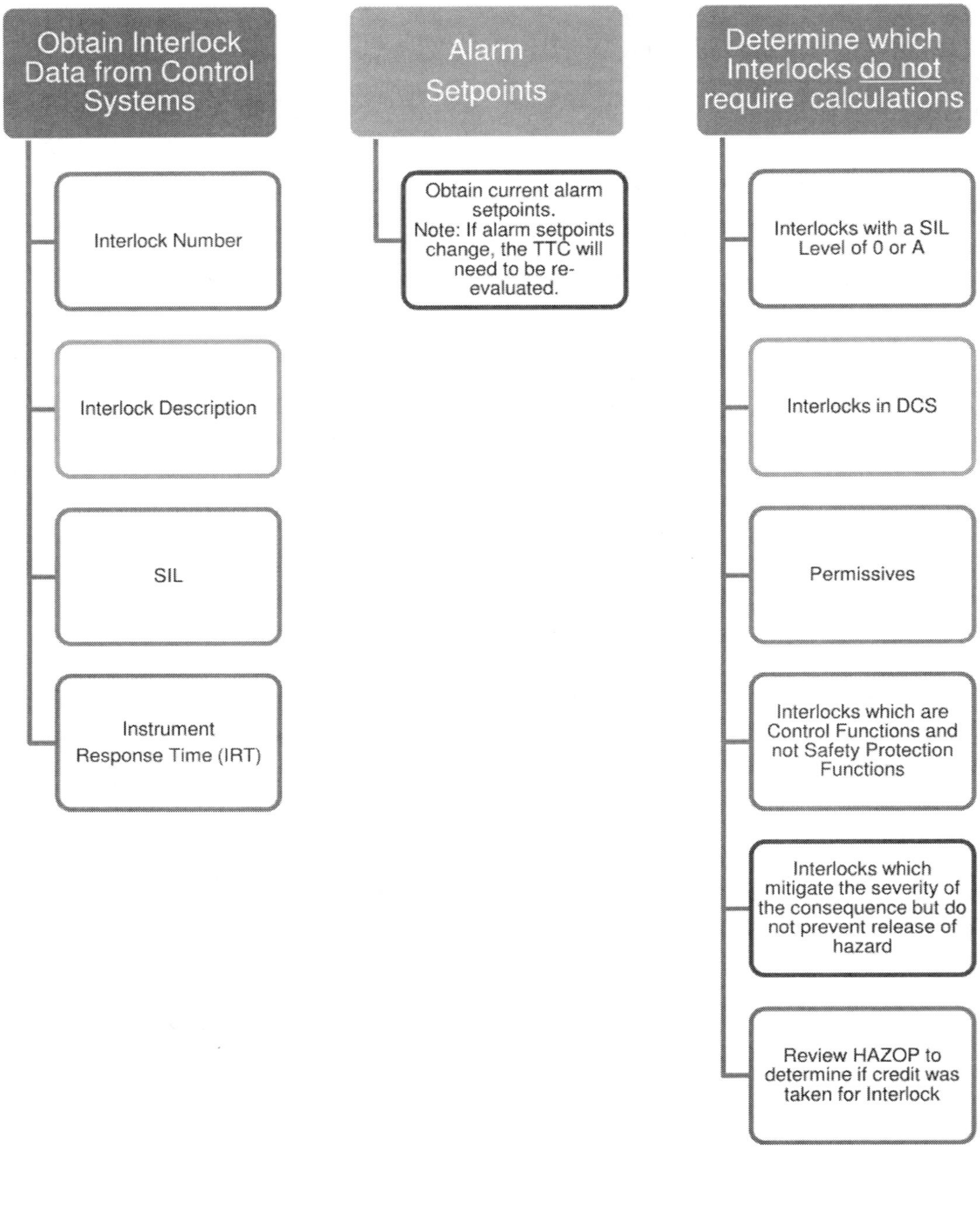

5.0 Methodology

The following summarizes the methodology of Time to Consequence calculations:
A more conservative approach of selecting SIF initiation point (Trip Setpoint) as the start point for the TTC is being followed at Bechtel. This method provides a higher safety margin, while designing the Safety Instrumented systems.

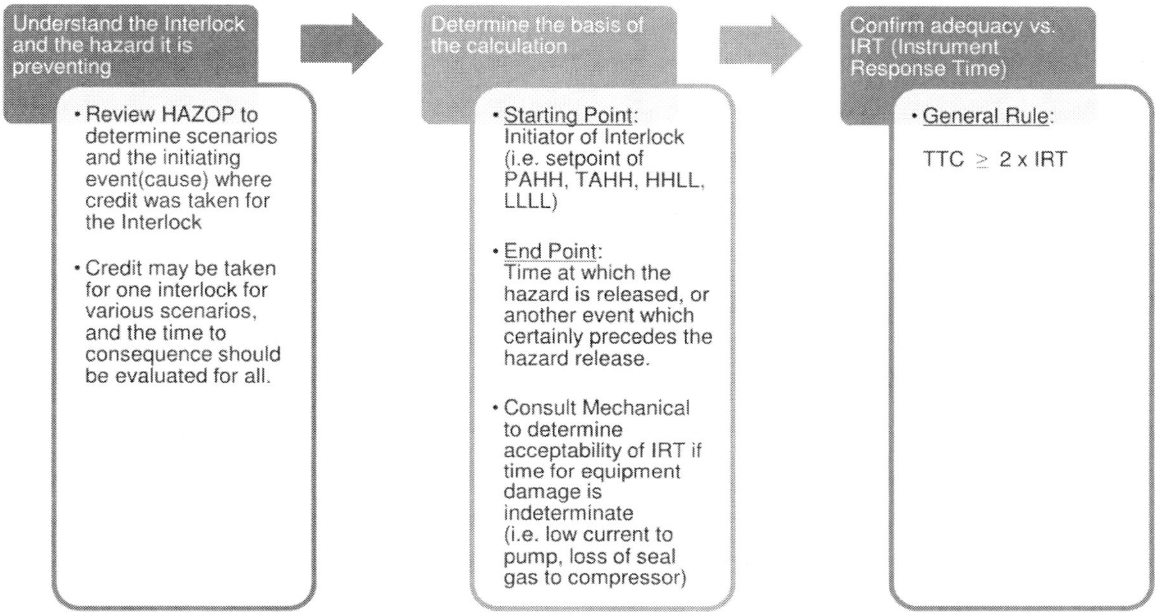

Insufficient TTC

The following can be looked at if the TTC is insufficient:

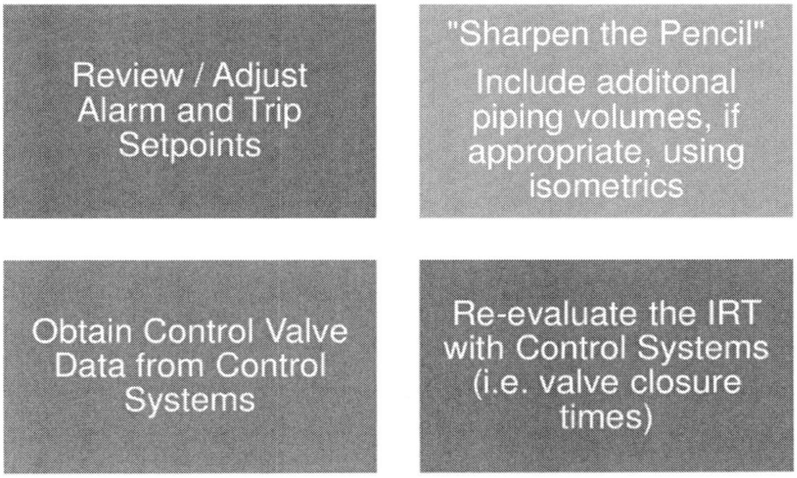

6.0 Case Study – TTC Calculations for LNG Facility

A Fail-Safe analysis was done for LNG Facility for all the safety interlocks to determine if the response of the Safety Instrumented System is quick and fast enough to protect the equipment or for containment. As a result of this, process engineers had to come up with Time to Consequence for different interlocks. Some calculation examples of different interlock are discussed below.

6.1 Calculation #1 - Simple Calculation.

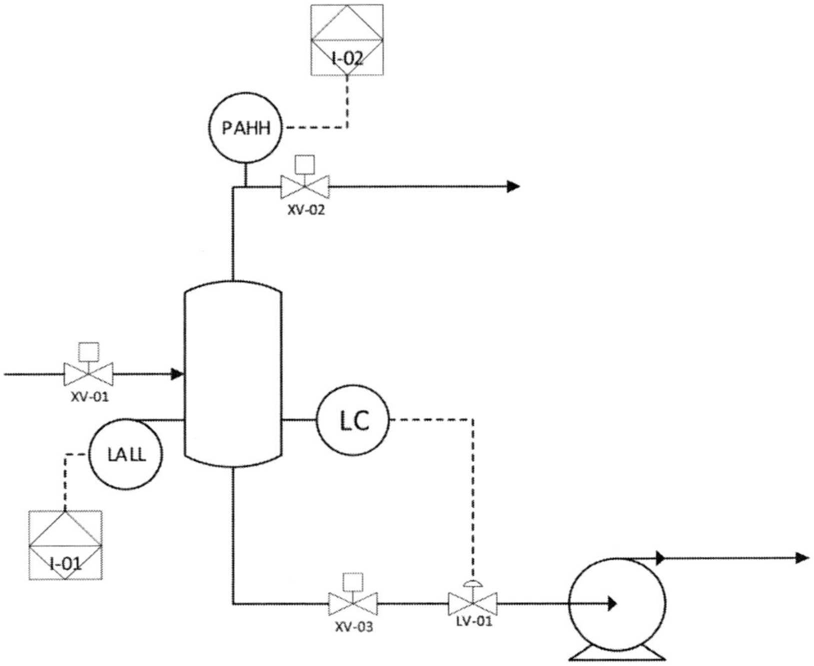

Figure 2: Flash Drum

Objective: Protect the downstream pump from loss of suction.
Interlock I-01 Description: Low-low level in vessel closes XV-03 on liquid outlet line.

Start point: LALL setpoint
End point: Vessel is empty.

Calculation:

Liquid volume from LLLL - bottom head = 20 m^3
Rated capacity of pump = 100 m^3/hr

Time to Consequence $\geq$ 720 seconds

6.2 Calculation #2 - Steady-State.

Objective: Protect the vessel from overpressure.
Scenario: XV-02 spuriously closes while feed to vessel continues.
Interlock I-02 Description: High high pressure closes XV-01 on inlet line to vessel.

Start point: PAHH setpoint.
End point: 110% of system design pressure.

Note: The pressure at which a vessel ruptures is the defined endpoint but is difficult to calculate. In these situations, the TTC calculation may be based on allowable accumulation over system design pressure instead.

Calculation,

Vessel vapor volume =	50 m3 (above HHLL, including top head)
PAHH set pressure =	10 barg
Design pressure =	13 barg
110% design pressure =	14.3 barg (10% accumulation, system is pneumatically tested)
Temperature =	60°C
R, gas constant =	0.0831 m3(bar)/K-kmol

$\Delta PV = \Delta nRT$ Assume ideal gas, Z = 1

Vapor feed rate = 500 kmol/hr

n at PAHH =	18.1 kmol
n at 110% design P =	25.8 kmol
Δn =	7.8 kmol

$$Time\ to\ Consequence \geq \frac{\Delta n}{vapor\ rate}$$

Time to Consequence ≥ 56 seconds

Additional considerations if TTC is not greater than the IRT:

- Evaluate if PAHH setpoint may be lowered while maintaining satisfactory margin above normal operating pressure to avoid spurious activation of interlock.
- If vessel volume is not sufficient, consider including piping volume up to isolation valves closed by the interlock based on isometrics.
- If wide open control valve is the source of overpressure, obtain the wide open flow rate data from vendor / Control Systems. Dynamic simulation may also be used to evaluate two-phase flow or vapor expansion if the upstream liquid volume exceeds the vapor volume in the downstream vessel.
- Review valve closure times with Control Systems.

6.3 Dynamic Analysis of Refrigerant Compressor System to Determine TTC.

Some scenarios require dynamic analysis to determine the time to consequence. Figure 3 shows a typical refrigerant compressor system driven by a gas turbine.

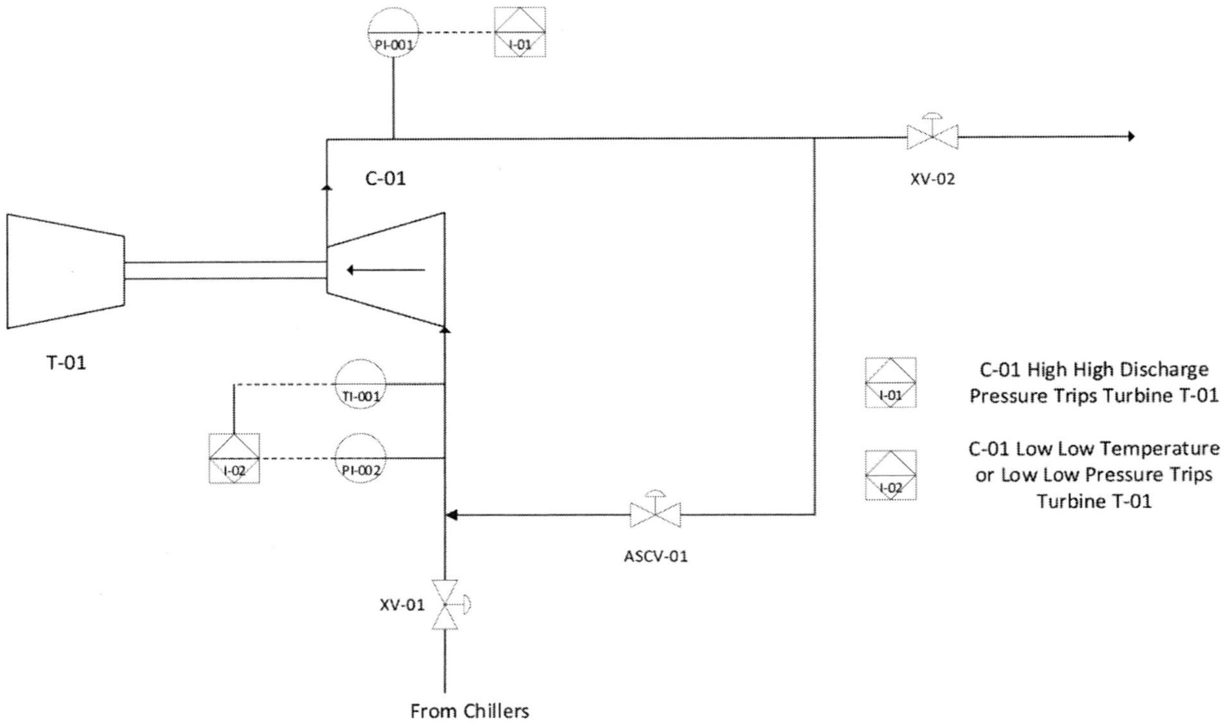

Figure 3: Refrigerant Compressor System

Determination of Interlock I-01 TTC

Objective

To determine if the High High pressure trip on a LNG refrigeration turbine will protect the system from overpressure in the event of a blocked discharge.

Interlock I-01 Description

Interlock I-01 is activated by high high pressure (PAHH-001) of refrigerant compressor C-01 during normal operation/startup. The trip set point of PAHH-001 is 21 barg. HS discharge piping design pressure is 23 barg.

Scenario

C-01 (XV-02) fails closed

Assumptions

The following assumptions were made for the dynamic runs to simulate this scenario.

1. Upstream feed gas flow rate is at normal operating conditions for the scenario when discharge valve fails closed.
2. Anti-surge control of all stages does not take action and remains in closed position. This is more conservative as the safety time with the anti-surge valve opening is more than with the valve remaining closed.
3. Discharge PSV does not lift.

Results

During this scenario (discharge isolation valves XV-02 actuator failed), discharge pressure increases beyond 110% of the design pressure (25.3 barg) as the anti-surge controllers are not active.

The time from high high pressure (PAHH-001) to 110 % of the discharge piping design pressure (25.3 barg) is 30 seconds. High Stage of C-01 surges immediately after discharge valve failure (~after 10 seconds) and the prediction of compressor performance at surge may not be exact. Surges are definitely detrimental to compressor but it is not the end point of rupture for the discharge piping.

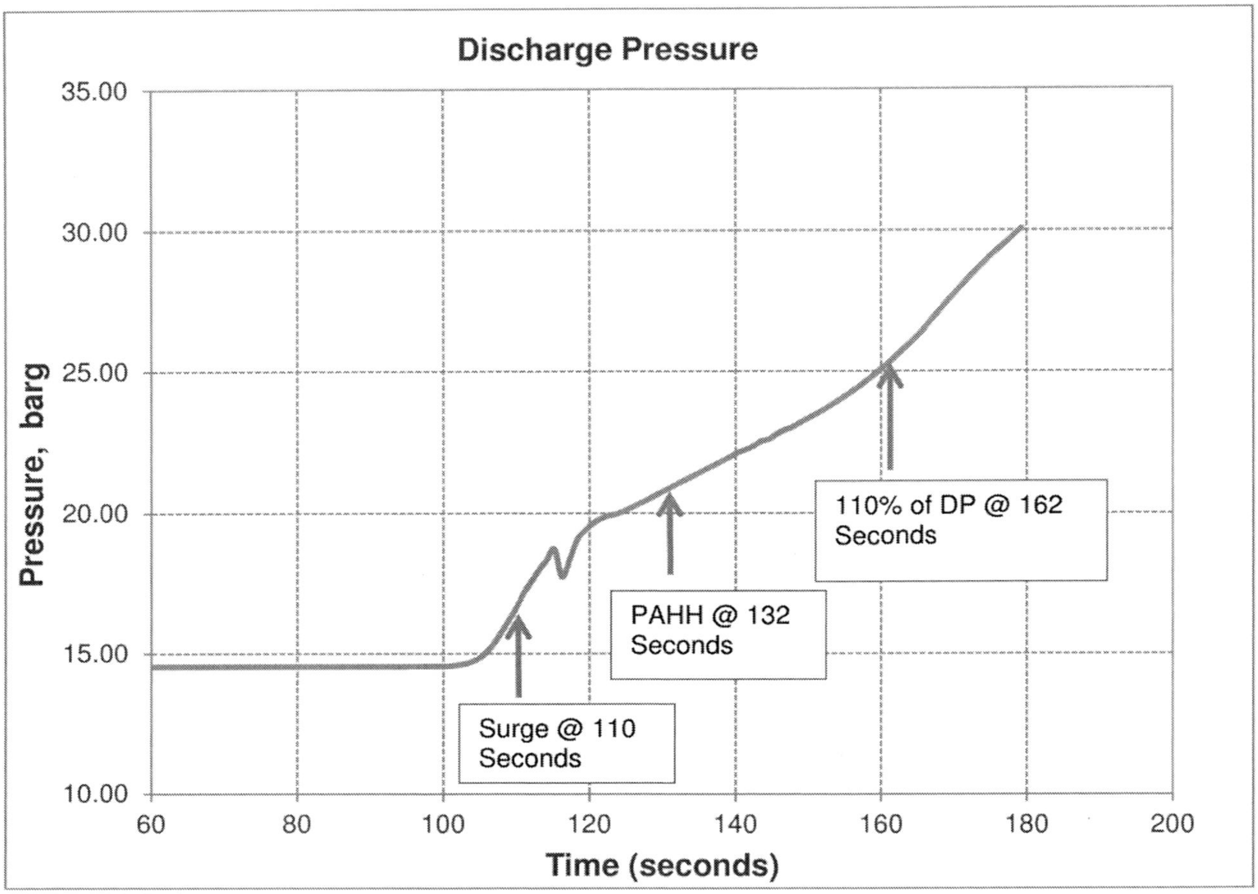

Figure 4: Discharge Pressure of C-001

An additional simulation was run with anti-surge controllers active, and in this case, the discharge pressure did not reach 110 % of the discharge piping design pressure within a reasonable time period.

The TTC for this high high pressure scenario is $\geq$ 30 seconds.

Determination of Interlock I-02 TTC

Objective

To determines the metal temperature cooling time from operating temperatures to MDMT for the compressor suction line.

Interlock I-02 Description

Interlock I-02 is activated by low low pressure (PALL-002) and low low temperature (TALL-001) of Refrigeration Compressor C-01 during normal operation/startup.

Methodology and Assumption

Pipe model in VMGSim Dynamics was used to determine the time for the compressor main suction line to reach the minimum design temperature specification (-40°C in this example). The starting point pipe metal temperature for the suction line is the fluid temperature (-35°C), which is based on the trip point temperature of TALL-001. The conservative decrease in process fluid temperature is the dew point temperature coincident with normal operating pressure.

Pipe Model Inputs

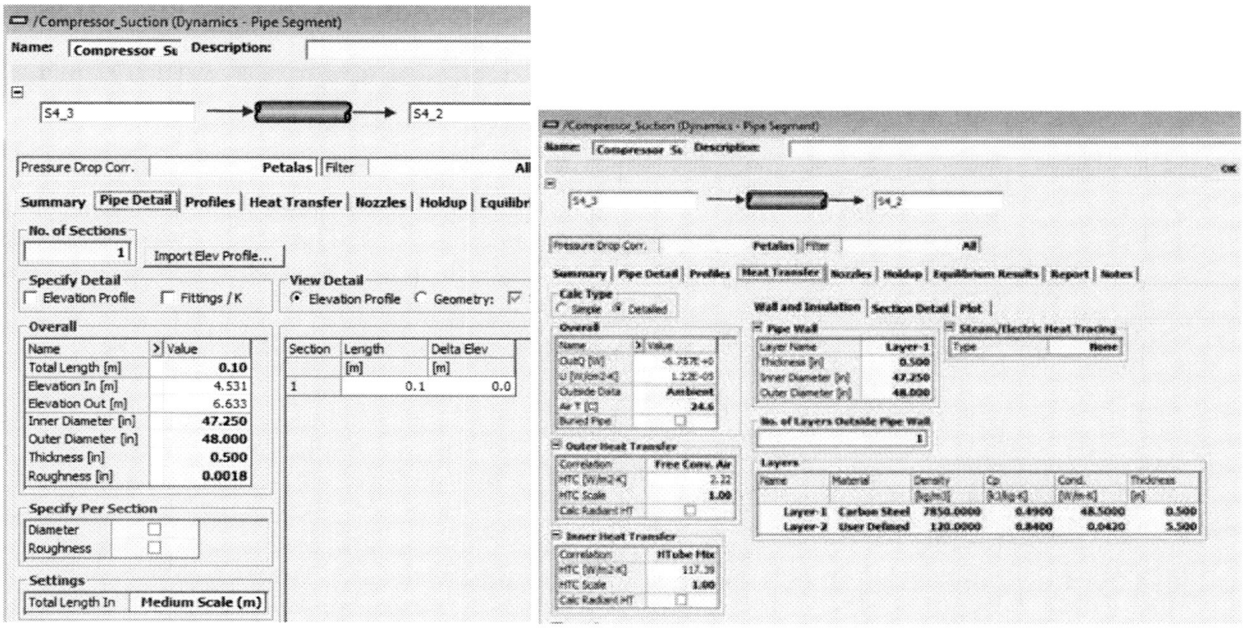

Figure 5: Pipe Model in VMGSim Dynamics

The above figure represents the data used to model a pipe segment. The pipe increment of 0.1m represents the dissimilar weld of the two different materials. The corresponding material physical properties such as density, Cp, and conductivity are the major inputs for heat transfer modeling in a pipe segment.

Results

The result from the pipe model shows how fast the first segment of pipe would take to reach -40°C from a metal temperature of -35°C when process fluid passes through with a worse case temperature of -90°C (dew point temperature).

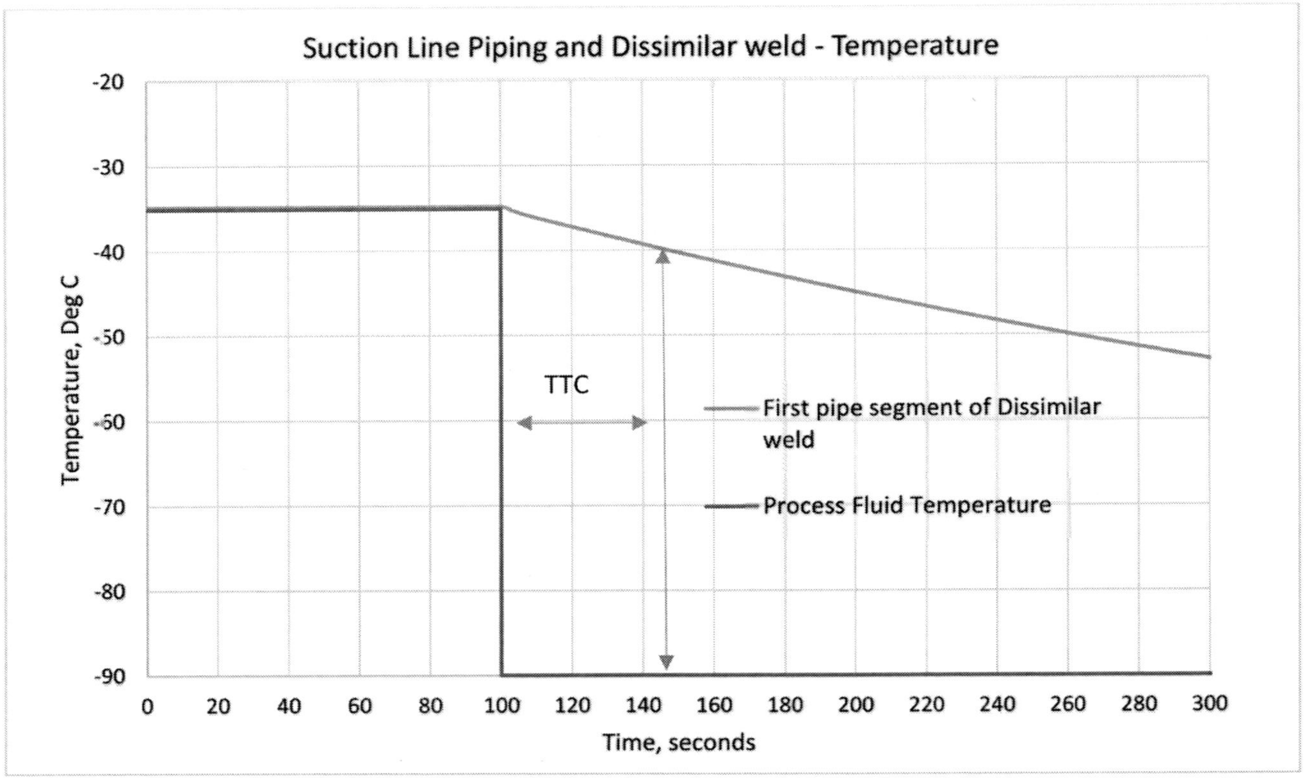

Figure 6: Suction pipe – Temperature.

The TTC for Low Low temperature scenario is ≥ 45 seconds.

7.0 Conclusion

Time to Consequence or Process Safety Time is not specified but rather estimated, calculated or measured (if applicable). TTC is system dependent and hence relates to the behavior of the process and process equipment or process control system within the context of a specific unmitigated hazard scenario. In this paper, the definition of TTC has been explained in detail with examples and the industry standards influencing TTC have been highlighted. Steady state calculation and dynamic simulation to determine TTC of complex system have been explained.

550577 Rational Screening of Adsorbents for Natural Gas Upgrading By Pressure Swing Adsorption Using Dynamic Simulation of Process Performance Metrics

The continuing development of a wide variety of novel adsorbents raises new possibilities for improving efficiencies in natural gas purification by pressure swing adsorption (PSA). However, designing an effective industrial scale PSA process requires expensive and time consuming pilot-scale testing. This highlights the need for effective *in silico* material screening tools. Currently, most performance metrics used for material screening focus on material properties such as adsorption capacity and product selectivity. However, restricting attentions solely onto the material properties effectively ignores critical features of the overall adsorption processes in PSA. On the other hand, studies that consider the overall process require the solution of complex design optimization problems for PSA given each adsorbent; This requires significant expertise from the practitioner. To address these limitations, a simplified dynamic model is presented along with heuristics based PSA process design rules that enable practitioners to easily evaluate relevant PSA process performance metrics through dynamic simulation, using only adsorbent material properties including isotherms and basic design specifications such as the feed composition and desired purity and recovery.

550354 Shortcut Modeling of Natural Gas Supersonic Separation

Supersonic separation is a novel technology for natural gas separation. The theoretical design uniquely combines concepts from aerodynamics, thermodynamics, physical separation and fluid-dynamics resulting in an innovative gas conditioning process. It is used to condition the gas by removing condensable vapors and natural gas liquids. The supersonic separator is composed of a converging section, a Laval nozzle and a diverging section. Natural gas flows from reservoirs with low velocity and high pressure. In the supersonic separation process, the temperature drops below the dew point of the natural gas. A multiphase flow is formed. Undesired components form liquid condensates that are centrifugally removed through side collection streams.

In this work, a one-dimensional numerical model was developed. The model presents a great potential as a fast and accurate tool that enables the simulation of supersonic separators with significant details. The model is comprised of rigorous thermodynamic and phase equilibrium calculations as well as multi-phase sound speed calculations. Although the Peng-Robinson equation of state was used, the approach is general and compatible with other equations of state. The model fills certain gaps found in literature with a shortcut modeling technique. It would best fit the category of preliminary design tools with significantly lower computational loads. Significance associated with the shortcut model pertains to its ability to simulate supersonic separators at a fraction of the time taken by other existing numerical models. Although the results will not detail the profiles of the thermodynamic conditions along the nozzle, six fundamental points are accurately predicted. Those points represent the inlet, nozzle throat, side stream location, upstream of the shockwave, downstream of the shockwave and the nozzle exit. These fundamental points serve as the necessary pivotal information in the conceptual design of supersonic separators.

Validation tests included different working fluids, feed conditions, nozzle geometries, specifications for side streams and liquid phase recovery specifications. Reduction in simulation run time and computational load represented by number of locations tested in the diverging section was demonstrated. A decrease of 75%-97%, depending on the specific case, in either the simulation time, number of locations tested or both has been achieved.

550873 Natural Gas Decomposition and Separation of Solid Carbon and Hydrogen Products in High Temperature Molten Salts

The lowest cost means of producing hydrogen and/or dispatchable electricity without CO_2 emissions in the U.S. is potentially through a process utilizing methane pyrolysis of abundant, low-cost, natural gas. The challenge of achieving high methane reaction rates and high conversion to molecular hydrogen and separable solid carbon may be met using high temperature (~1000 ^oC) complex molten salt mixtures together with catalysts. High rates (>1 mole/m^3-s) of methane decomposition and high selectivity for molecular hydrogen and solid carbon have been achieved with several multiphase salt-catalyst systems. When the physical properties of the liquids are selected optimally, separable solid carbon produced from pyrolysis in the high temperature melts which may be conveniently separated in bubble column reactors. In molten metal alloys and/or mixtures of irreducible halide salts hydrogen and graphitic carbon products are produced at high conversion and high selectivity and in forms which are separable from the melt in a single reactive separation. Solid catalysts (including Ni/Al_2O_3) suspended in specific melts are also shown to be continuously reactivated as the high temperature liquid serves as a solvent to remove carbonaceous surface deposits prior to the irreversible formation of deactivating coke. Single pass methane conversion of over 98% to molecular H_2 at over 98% selectivity is demonstrated in several complex melt systems. Fundamentals of C-H bond activation on melt surfaces and the interactions with the gas phase chemistry will be discussed and distinctions made as to how the subsequent reaction pathways can produce different pyrolysis pathways and different types of solid carbons. Process designs for solid carbon synthesis with zero carbon hydrogen and/or electricity production together with relative technoeconomics will be presented.

550324 Unique Applications for Embedded Model Predictive Control Technology

Some modern Distributed Control Systems have Model Predictive Control (MPC) technology embedded into the control system. The MPC is intended for small to medium MPC applications and, due to low implementation cost, have a good return on investment for small applications. Benefits of being embedded in the control system include use of the standard configuration application, standard operator interfaces and ease of implementation. In some systems, the embedded MPC can execute as fast as 1 per second. These attributes make the embedded MPC technology suitable for many applications that are considered difficult or complex control with PID controllers. These include such applications as a dead time dominate loop, override applications, feed forward, "small valve-large valve applications" and interacting loops. This presentation will show the use of embedded MPC technology for small applications that have an excellent return on investment.

550612 Linear Control in a Nonlinear World (A Practical Perspective)

Most process control applications assume that the processes they control are "linear". In other words, they should respond the same way at high or low rates, when moving in one direction or the other, and when incremental changes are small or large. Unfortunately, most "real world" processes aren't so simple and exhibit nonlinear behavior in different ways and to varying degrees. In this presentation, we will review some common nonlinear process characteristics and the differences between traditional linear and true nonlinear control (NLC) applications. We will discuss various linearization techniques that are used to improve the performance of linear control applications, including their relative difficulty/complexity, potential benefits, and limitations.

550393 Shifting Real Time Optimization Tasks to Plant-Wide APC Layer By Introducing Unconstrained and Constrained Optimum Concepts

Shifting Real Time Optimization tasks to Plant-wide APC Layer by Introducing Unconstrained and Constrained Optimum Concepts

James Wu

LyondellBasell Industries

Abstract to be submitted to the process control and optimization session in 2019 Spring AIChE meeting

Plant-wide optimization plays a critical role in improving chemical plant and refinery economic performance. Traditionally, steady state rigorous modeling optimization approaches have been used. There are several key challenges, such as using steady state models to deal with dynamic operating conditions and constraints, coordination with APC layer, etc. See a comprehensive RTO review by M.L. Darby et al. [1]. In the recent decades, "grey-box" nonlinear dynamic optimizers have been implemented to replace steady state optimizers. In addition, attempts have been made to move optimization functions to APC layers. One popular approach widely used in olefin plants is to use composite linear programming to coordinate plant-wide APC applications and to push feed against capacity constraints. But, it cannot be used for non-linear yield-related optimization trade-offs[1].

In this paper, a new plant-wide optimization approach is developed for those plants where individual APC applications can be integrated under a composite linear programing application or all key optimization trade-offs are included in one large APC application. It shifts daily tasks related to dynamic optimization trade-offs under constrained conditions from the RTO layer to the plant-wide APC layer. This is done by introducing the concepts of unconstrained and constrained optima. It shows that for many practical applications unconstrained and constrained optima defined here depend mainly on pricing and feed composition changes, but not on real time operating conditions. As a result, the plant-wide APC layer can be used for daily real time optimization based on unconstrained and constrained optima. These optima can be updated using either existing steady state optimizers (with modification and much reduced frequency) or rigorous offline process models. This new approach can significantly improve optimization performance under dynamic capacity constraints, and can also reduce overall optimization implementation and support efforts.

[1] Mark L. Darby, Michael Nikolaou, James Jones, Doug Nicholson, RTO: An overview and assessment of current practice, J. of Process Control 21 (2011) 874-884.

Executive Summary

Title: Open Process Automation: Unlocking Value Capture with Next Generation DCS

Authors: Bradley G. Houk, David L. DeBari, Tyler A. Soderstrom; ExxonMobil Research and Engineering

Meeting: American Institute of Chemicals Engineering 2019 Spring Meeting, Fuels and Petrochemicals Division, Process Control and Optimization Developments

Location and Date: New Orleans, LA. April 3, 2019

Distributed Control Systems (DCS) are critical for the safe and reliable operation of today's refineries and petrochemical plants. The DCS also provides the foundation for our advanced process control and real-time optimization applications. Over the years, these applications have become a key component for maximizing the value from our manufacturing plants. However, the traditional closed proprietary industrial control systems have a number of challenges. ExxonMobil is working in conjunction with the Open Process Automation™ Forum (OPAF) of The Open Group which includes representatives from a diverse group of industries to create an alternative.

The goal of the activity is to enable the creation of a standards-based, open, secure and interoperable process control architecture. This will enable next generation process control systems to be built from commercially available components and be used by a wide range of operating companies across many industries. An Open Process Automation (OPA) system will support integration of best-in-class components and provide both configuration and application portability across components from different suppliers.

A move from the traditional closed proprietary industrial control systems to an open architecture based upon industry standards is enabled by several factors. The use of industry standards to create interoperable systems based upon commodity hardware has been achieved in other industries. The greater integration of IT and OT is opening up a range of new possibilities. Finally, end user expectations have grown as computing power continues to grow and with the increased use of big data and data analytics.

Success depends on active engagement by end user companies, suppliers, and system integrators to select the standards. This broad industry representation is achieved under the leadership of the Open Process Automation™ Forum (OPAF) of The Open Group. The OPAF currently has 88 member companies including 22 end user companies.

In parallel to the standards work by the OPAF members, ExxonMobil is executing an internal R&D program to demonstrate the art of the possibility. In 2018, ExxonMobil completed the work on an OPA proof of concept system which demonstrated key quality attributes using a heterogeneous system containing components from 10 different suppliers running against a simulated fired heater. ExxonMobil is currently developing an OPA prototype system which will be deployed to control a pilot scale research unit. A multi-month demonstration run with the OPA prototype will provide the first time use in hydrocarbon service. Information from the ExxonMobil R&D activities are shared with OPAF to help guide the selection of appropriate standards.

Six operating companies have joined ExxonMobil to execute a Collaboration Agreement to advance the development of OPA systems. The companies are Aramco Service Company, BASF, ConocoPhillips, Dow,

Georgia Pacific and Praxair. The ultimate goal of the collaboration activities is for each company to create independent field trials of OPA systems built from commercially available products.

The move to standards-based, open, secure and interoperable industrial control systems will provide a range of benefits. Next generation process control systems should support reuse of applications across different supplier platforms. Value capture will also be supported by enabling the deployment of independently developed software and breaking the dependency between a supplier's software and hardware. Security will be designed in from the beginning and incorporate the latest technology to address the growing cybersecurity threats.

In summary, the time has arrived to explore moving from the traditional closed proprietary industrial control systems to an open architecture based upon industry standards. A growing number of end user companies, suppliers and system integrators are actively engaged in selecting standards. Companies are collaborating to accelerate the development of field trials, and driving the need for commercially available components. The critical mass of end users to create demand and suppliers to bring products to the market are keys for success.

546694 The 8 Rules for Successful Controls Implementations

A project team for an overseas grassroots LNG plant struggled for months attempting to develop a comprehensive series of control schemes for a vital plant utility unit. Having no success, the EPC and DCS Vendor eventually turned to an experienced Process Control Engineering Consultant for help. In a matter of a few days, a conceptual design was developed and approved, followed by detailed design and testing within a virtual environment. The design was then successfully implemented at site, along with numerous enhancements to address unexpected instrumentation and process design issues.

Upon completion of commissioning, the utility systems, as well as the balance of the plant, was run in fully automatic control – something that had not been experienced since the plant was originally commissioned.

Using this real-world situation as a backdrop, this presentation illustrates the practical application of the "8 Rules" and how it provided Operations with the handles and situational awareness they needed to reliably control the plant. It also illustrates how a creative solution avoided the main issue that stymied the project team: a documented DCS "feature".

545134 A New Look at Industrial Incidents *Driving Operating Discipline in the Control Environment*

A New Look at Industrial Incidents

Driving Operating Discipline in the Control Environment

By Tom Nolan, Senior Process Control Consultant, Emerson

Industrial incidents injure people and cause millions of dollars in damages. Several critical shortcomings have been identified as significant contributors in past incidents, and yet these incidents are still happening today—often resulting from the same contributing factors. Though incidents typically coalesce from several issues, there are common factors that can and should be managed. By leveraging dynamic alarm management and operating discipline in a state-based control environment, organizations can strike the necessary balance between human operation and automation support to enhance operability and improve safety performance.

551076 Minimizing CO_2 production during Syngas Catalysis Mediated By Nano-Sized Iron Supported on Silica Spheres

Minimizing CO_2 prodcution in chemical reactions is a desirable pathway to low-carbon processes. Our approach encompasses two aspects: 1) use nano-sized catalysts to enhance surface area to achieve increased space-time-yield (STY) and couple it with 2) slurry phase for better heat management. For this study, we selected well-known iron mediated Fischer-Tropsch (F-T) synthesis as the test reaction due to issues with product selectivity and catalyst activity and the fact that CO_2 is a major side product in conventional F-T reaction. In the study, three a-Fe_2O_3-based catalyst precursors were evlauted: 1) an unsupported nano-sized (20-80 nm) BASF material, 2) a micrometer-sized (32.5 μm) UCI supported on gamma alumina and doped with potassium as a reference, 3) a core-shell configuration consisting of silica microspeheres (500 nm) as the core and about 20 nm iron oxide as the shell. All three materials were evaluated in a 300 mL Parr batch reactor for F-T synthesis at three temperastures with the goal to shut-sown water-gas-shift (WGS) reaction that is the cause of CO_2 prodcution. The F-T data were collected by feeding synthesis gas (syngas) with H_2/CO ratio of 2.

The presentation will first focus on key steps of the core-shell nano-sized material synthesis and its characterization using a suite of spectroscopic techniques including Transmission Electron Microscopy and Raman. Then, its perfornace under F-T synthesis reaction will be discussed for CO_2 minimization relative to desired hydrocarbon products, thus enhancing overall carbon conversion. We will present morphology and structural characterization data and full mass balance by analyzing gas and liquid products of each run. A relationship between structural properties and reaction parameters to achieve low CO_2 or non-CO_2 production scenarios will be discussed.

COMPARISON OF THE EQUILIBRIUM CONSTANT OF THE WATER GAS SHIFT (WGS) REACTION FOR DETERMINATION OF REACTION CONVERSION

Vagner dos Santos, Faculdade Estácio, Sergipe, BR

ABSTRATC

Oil is a major source energy and has a strong impact on environment pollution by the gases emitted in the burning of its derivatives, for example, carbon monoxide, carbon dioxide. A way of reducing the emission of such gases as for example CO is converting by means of reaction in products less aggressive to the environment as is the case of the hydrogen gas. The reaction *water-gas-shift* is a secondary reaction that produce hydrogen and carbon dioxide from the reaction carbon monoxide with water, this reaction occurs in the synthesis of Fischer-Tropsch, in the synthesis methanol, conversion of hydrocarbons by reform among others. The equilibrium composition of the reaction can be estimated from the reaction equilibrium constant under various temperature, pressure and feed composition conditions. The equilibrium constant can be estimated from thermodynamic properties such as enthalpy and entropy, the constant suffers great influence as temperature. In the literature there are some correlations to determine the equilibrium constant of the WGS reaction. For the modeling was considered real system, the reaction occurs in the gas phase, the thermodynamic equilibrium occurs, the coefficient of fugacity and the Soave/Redlich/Kwong equation of state were used. The whole simulation was developed in the Scilab, which obtained the equilibrium constant with the models found in the literature and compared with data of equilibrium constant also found in the literature, for the model that presented the best result was used for the simulations, the chosen conditions were for several temperature and feed conditions (CO/H_2O). The results obtained were the conversion versus CO/H_2O ratio for various temperatures, the conversion versus temperature for several CO/H_2O ratio and the formed hydrogen composition versus the CO/H_2O ratio for various temperatures.

Introduction

Petroleum is still the most widely used energy source in all countries, and also the main source of raw material for the petrochemical industries responsible for the production of various products used in our daily lives, such as plastics, rubber, cosmetics and products cleaning.

The search for alternative renewable energy sources is something that universities and research centers are looking for, as oil is not a renewable source of energy, because the oil is extracted from the fields that go through refining processes to produce various types of fuels such as: gasoline, diesel, kerosene and others. However, these fuels are burned, and on burning are released into atmosphere gases such as carbon monoxide, carbon dioxide and sulfur oxides which are responsible for increasing the greenhouse effect and pollution of atmospheric air.

One way to minimize the effects of these gases in the atmosphere is to convert them into useful substances for our day to day, one of the useful substances are fuels, polymers among others, however, one way to convert these gases is by means of chemical reactions. The water-gas-shift (WGS) reaction is an alternative to reduce the emission of carbon monoxide (CO), the water-gas-shift (WGS) reaction is a reaction that reduces carbon monoxide in a rich stream in hydrogen gas (Y. Choi, HG Stenger, 2003).

Hydrogen is a source of clean energy and can be produced from hydrogen cells by the electrolysis process. Hydrogen is a fuel that is being explored more and more for being considered clean fuel. It is a raw material widely used in the industry in the production of various products such as ammonia, methanol and in process as in the reduction of iron ores, another branch is in the fuel cell as an alternative source of clean fuel (BESTETI, 2005). It is also used in hydrogenation reactions of unsaturated hydrocarbons, a process used in hydrotreatment units in oil refineries.

Water Reaction Gas Shift

The *water-gas-shift* (WGS) reaction is used to obtain hydrogen and carbon dioxide from the reaction of carbon monoxide and water as can be observed in the reaction below, it is also a reversible and exothermic reaction.

$$CO + H_2O \rightarrow CO_2 + H_2 \qquad \Delta H°_{298} = -41{,}2\ kJ/mol$$

From the point of view of thermodynamics, CO conversion is favored under low temperature conditions, causing the reaction to shift towards the products. Of the factors that can influence the thermodynamic equilibrium, the pressure is the factor that does not influence the equilibrium of the reaction, that is, the increase or the decrease of the pressure will not influence the formation of the products. By observing the stoichiometry of the WGS reaction, the number of moles of the reactants is equal to the quantity of moles of the products. The ratio of the reactants CO / H_2O is another factor that can influence the equilibrium of the reaction.

In the Fischer-Tropsch (FT) synthesis the water-gas-shift (WGS) reaction occurs simultaneously, however, the WGS reaction is an unfavorable reaction for FT synthesis, this is because the WGS reaction produces hydrogen which occurs the opposite in the FT synthesis which the hydrogen is consumed.

The *Water-gas-Shift* (WGS) reaction is not the main reaction because it is a reaction that occurs in a secondary way simultaneously with the following reactions, methanol synthesis, conversion of hydrocarbons by reforming processes, Fischer-Tropsch synthesis, exhaust gases from cars passing through the catalyst and the selectivity of carbon monoxide oxidation in fuel cells (Ladebeck, Wagner, 2003).

Industrially the WGS reaction occurs in two stages to achieve a greater conversion of carbon monoxide, one at high temperatures and others at low temperatures, and to overcome the thermodynamic limitation at low temperature conditions. At high temperature (HT) the working temperature ranges from 320 to 450 ° C. However, the reaction achieves a high conversion of CO in the low temperature stage, which occurs in the range of 150 to 250°C (Newsome, 1980; Grenoble *et al.*, 1981).

Equilibrium constant

The equilibrium constant of a feed is the ratio between the fugacity coefficients of the products and the reactants. The equilibrium constant can be determined from the thermodynamic properties of the reactants and the products such as enthalpy and entropy of each species composing the reaction and stoichiometry. The literature presents several works that use the equilibrium constant for several reactions and application. From the equilibrium constant can estimate the composition of the substances that make up the reaction system when the thermodynamic equilibrium under temperature and pressure conditions is reached.

In the work done by P.L. Mores *et al* (2018), he estimated the composition of the gases involved in the combustion reaction that occurred in a power plant for CO_2 sequestration. For the estimation of the combustion gases, the equilibrium constant of the combustion reaction was used. Newsome (1980) and Byron *et al.* (2010) in their work to study the *water-gas-shift* reaction rate (WGS) used correlations that estimated the equilibrium constant.

Rahimpour (2011) in his studies on hydrogen production, he simulated the process to produce hydrogen gas from the natural gas processing, where a hydrogen selective membrane was used, for this was considered the WGS reaction, the equilibrium constant used in modeling was the constant proposed by Wang *et al* (2003).

Modeling and simulation of equilibrium constants

The correlations used to estimate the equilibrium constant were obtained in several studies in the literature, from the correlations the simulations were performed and compared with data from the literature.

Newsome (1980) in his paper used the correlation shown below to estimate the equilibrium constant.

$$K_{eq} = exp\left(\frac{4577,8}{T} - 4,331\right) \tag{1}$$

Byron *et al.* (2010) used the correlation to estimate the equilibrium constant of the reaction, the correlation is shown by equation 2.

$$\ln K_{eq} = \frac{5693,5}{T} + 1,077x \ln T + 5,44.10^{-4}.T - 1,125.10^{-7}.T^2 - \frac{49170}{T^2} - 13,148 \tag{2}$$

Wang *et al* (2003) proposed a correlation of the equilibrium constant that was used by Rahimpour (2011) in his research on the reaction and water-gas-shift. Correlation is represented by equation 3.

$$\ln K_{WGS} = \frac{5078,0045}{T} - 5.8972089 + 13.958689 \times 10^{-4}T - 27.592844 \times 10^{-8}T^2 \tag{3}$$

Resultados

The keywords found in correlations by Byron (2010), Newsome (1980) and Wang *et. al.* (2003) are compared with the data found in Süd-Chemie.

Table 1 - Comparison of equilibrium constant models

Temperatura (°C)	Süd-Chemie	Byron (2010)	Newsome (1980)	Wang *et al* (2003)
93,3	4523	5247,2	3502,705	5247,19
148,9	783,6	885,56	675,59	885,56
204,4	206,8	229,63	191,52	229,63
260	72,75	79,71	70,48	79,70
315,6	31,44	34,16	31,326	34,20
371,1	15,89	17,13	16,032	17,13
426,7	9,03	9,663	9,117	9,658
482,2	5,61	5,971	5,637	5,975
537,8	3,749	3,975	3,721	3,971
593,3	2,653	2,800	2,592	2,802
648,9	1,966	2,070	1,88	2,069
704,4	1,512	1,594	1,422	1,591

It is observed that the equilibrium constants decrease as the temperature increases accordingly the conversion of carbon monoxide (CO) to hydrogen (H_2) is lower at higher temperatures.

Comparing the equilibrium constants obtained by the correlations with the constants found in Süd-Chemie, it is observed that the constants present results in the same order of magnitude, however, with a difference between the constants at low temperature, however, as the temperature increases difference between the constants get smaller and smaller.

As the values of the equilibrium constant presented close values between them, the correlation of Newsome (1980) was chosen to simulate the conversion of the carbon monoxide as a function of the CO/H_2O ratio at different temperatures. The result of the simulation is shown in the figure below.

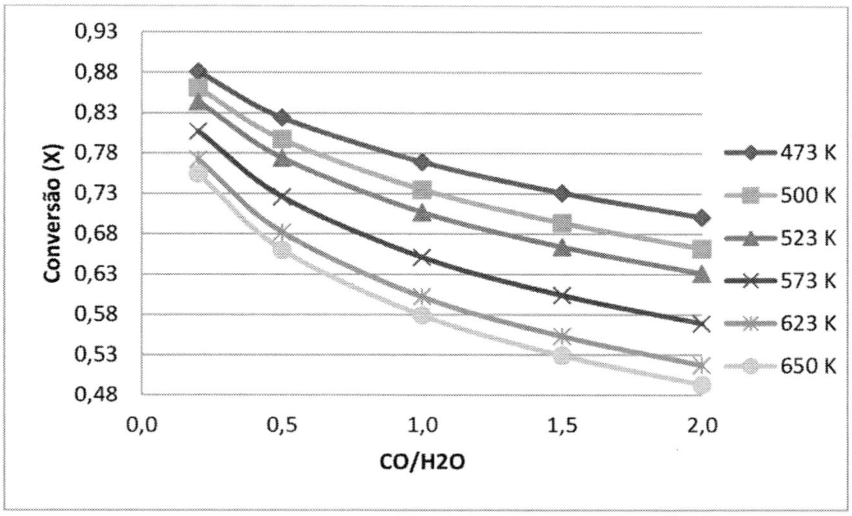

Figure 1: conversion of carbon monoxide to CO/H_2O

The WGS reaction is slightly exothermic, that is, as the temperature increases the conversion decreases, because the equilibrium of the reaction is displaced to the opposite direction of the reaction, favoring the formation of the reactants, this behavior is represented in figure 1 for several values of the CO/H_2O ratio.

The feed composition has a strong influence on the conversion of the reaction. For high CO/H_2O ratio it is possible to observe that the conversion of carbon monoxide decreases. Carbon monoxide is the limiting reactant of the reaction, whereas water is the excess reactant, however, when the ratio of reactants increases the amount of carbon monoxide is higher, but not all the CO is converted.

Conclusion

The results obtained in the simulation presented coherence with equilibrium constants found in the literature, even with small differences when compared with Süd-Chemie. All the correlations shown in this article can be used for simulations, but the correlation used by Newsome (1980) is the most appropriate to be implemented in algorithms, it is a correlation that presented good results in the simulation and because it is a simple correlation, its implementation will require less effort from the computer to do the calculations and presents the lowest risk of the programmer going wrong when writing command lines in programming.

References

1. AHÓN, V.R., COSTA Jr, E.F., MONTEAGUDO, J.E.P., FONTES, C.E., BISCAIA Jr, E.C., LAGE, P.L.C; A comprehensive mathematical model for the Fischer-Tropsch synthesis in well-mixed slurry reactors, Chemical Engineering Science, 60 (2005) 677-694.
2. BESTETI, M. D., SANTOS, D. G. F., CARVALHO, S. C., ASSIS, J. A. Avaliação Termodinâmica da produção de Hidrogênio A partir da Reforma a Vapor do Metano. VI Congresso Brasileiro de Engenharia Química em Iniciação Cientifica, 2005.
3. BYRON, SMITH R.J. et al. Review of the water gas shift Reaction Kinetics. International Journal of Chemical Reactor Engineering, v. 8, 2010.
4. CHOI. Y, STENGER. H.G; Water gas shift reaction kinetics and reactor modeling for fuel cell grade hydrogen, Jounal of Power Sources, 124 (2003) 432-439.
5. Grenoble, D.C., M.M. Estadt, and D.F. Ollis, The chemistry and catalysis of the water gas shift reaction: Journal of Catalysis, 1981. 67(1): p. 90-102.
6. LADEBECK, J.R., WAGNER, J.P; Catalyst development for water-gas shift. Handbook of Fuel Cells – Fundamentals, Technology and Applications. Volume 3, Part 2, pp 190 – 201. 2003.
7. NEWSOME, D. S., The Water-Gas Shift reaction Reaction. Catalysis Reviews, Science and Engineering, v. 21, p. 275-318, 1980.
8. PATRICIA L. MORES, JUAN I. MANASSAILDI, NICOLÁS J. SCENNA, JOSE A. CABALLENO, MIGUEL C. MUSSATI, SERGIO F. MUSSATI. Optimization of the designe, operating conditions, and coupling configuration of conbined cycle Power plants and CO_2 capture processes by minimizing the mitigation cost. Chemical Engineering Journal 331 (2018) 870-894.
9. RAHIMPOUR, M.R, MIRVAKILI, A., PAYMOONI, K.; Simultaneous hydrogen production and utilization via coupling of Fischer-Tropsch synthesis and decalin dehydrogenation reactions in GTL technology, International Journal of Hydrogen Enegy, 36 (2011) 2992-3006.
10. RAHIMPOUR, M.R., GHORBANI, A., ASIAEE, A., SHARIATI, A. Conversion of refinery natural purge gases to liquid hydrocarbons in GTL loop with hydrogen-permselective membranes: An alternative to gas flaring. Journal of Natural Gas Science and Engineering 2(2011), 461-475.
11. Wang, Y.N., Ma, W.P., Lu, Y.J., Yang, J., Xu, Y.Y., Xiang, H.W., et al., 2003. Kinetics modeling of Fischer-Tropsch Synthesis over an industrial Fe-Cu-K catalyst. Fuel 82, 195-213.
12. Süd-Chemie, 'Physical and Thermodynamic Properties of Elements and Compounds', Technical Bulletin, Süd-Chemie Inc.

550550 Process Optimization of Drm+Cosorb Process for Syngas Production

Dry Reforming of Methane Process utilizes CO_2 and CH_4 to produce syngas (mixture of CO and H_2). This process has grabbed a lot of attention in academia due to its utilization of CO_2 to produce a valuable intermediate. However, the process suffers from many challenges, that have to be addressed before it can compete with existing commercial processes. The three prominent obstacles are : severe coking regime, high endothermic nature of reaction and low quality of syngas obtained ($H_2/CO \leq 1$). Recent work done in our group [1] identified that a DRM stand-alone unit followed by a syngas ratio adjustment step to remove CO (COSORB) is a potential pathway towards utilization of the DRM process. In the high-level optimization study, it was found that this process is competitive when compared to commercial processes (Auto-thermal Reformer, Partial Oxidation) with reductions in CO_2 emissions above ~60% and reduction in operating costs above ~20% (in certain specific scenarios as detailed in [1]). The final usage of the captured CO has a big effect on overall process economics and carbon footprint. Hence, finding a reliable CO consumer is crucial to the success of the proposed process.

In this work, we have used various process optimization techniques (ASPEN Plus Optimization, LINGO Optimization) to get the best process operating parameters for the DRM+COSORB process while maximizing syngas production and minimizing overall CO_2 emissions. Since this combination has not yet been studied in literature, there is room for process improvements. The ultimate goal is to synthesize an optimized process flow-sheet for the proposed DRM+COSORB process. A comparison between the optimized flow-sheet and current commercial processes will be made to show the technical and commercial prospects of the proposed process.

Keywords : Dry Reforming, Syngas Ratio, DRM, COSORB

References

[1] S. Afzal, D. Sengupta, A. Sarkar, M. M. El-Halwagi, and N. O. Elbashir, “Optimization Approach to the Reduction of CO_2 Emissions for Syngas Production Involving Dry Reforming,” *ACS Sustain. Chem. Eng.*, vol. 6, p. 7532–7544, 2018.

550435 Use of Nonlinear and Machine Learning Techniques for Improved APC Modeling

Objectives

Model identification for advanced process control (APC) applications including model predictive control (MPC) and soft-sensor (or inferential) is one of the most time-consuming steps which requires dedicated plant tests and the skills of a highly experienced chemical engineering control engineer. The empirical models (such as first-order plus dead time, or FOPDT) need to be validated by judging the model attributes such as the gain value or sign, the dead time, and the shape of the step response based on prior process knowledge [1]. Plant tests must be repeated when the model identification does not match with the prior process knowledge.

To reduce the effort and cost of the model identification, this study proposes a simultaneous model fitting method which incorporates a priori information as constraints of the optimization problem. This approach can be applied to both linear empirical models and non-linear first principles models using the GEKKO optimization suite which specializes in dynamic optimization problems including nonlinear programming (NLP), differential algebraic equations (DAE), and artificial neural networks (ANN) [2].

Case studies and discussions

Two model identification examples are considered in this study. One is developing the inferential model for a soft-sensor in a distillation column. The other is a horizontal tank with a level control loop. Soft sensors play a critical role in estimating product quality (i.e. composition) in a real-time manner, especially in the chemical processing industry. They infer the product quality in the distillation column based on measurable column operating variables, such as tray temperature and pressure. The low frequency and time delay of on-line analyzers and lab sample results are not suitable for the real-time control applications. In the inferential fitting case, the industrial distillation column operation data is used to estimate the model parameters of the inferential model. The model is structured with three different blocks in series: the pressure compensated temperature (PCT) block, the FOPDT block, and the static nonlinearity block. For the nonlinearity block, a piecewise linear (PWL) function and an ANN are used and compared with a real-world result that uses a simple log transformation. Introducing the PWL and ANN as a nonlinear static gain block describes the nonlinearity of the process more accurately in an extended range of operation, which is difficult to achieve with a simple logarithm transformation.

In the tank example, instead of having a chain of model blocks, a nonlinear physics-based model is considered. The model is designed to accurately describe the nonlinear character of a horizontal tank with spherical caps. It has an inlet and an outlet flow with a PID level control loop configured on the outlet flow. The physics-based model consists of material balance equations combined with the tank geometry. The unknown parameters (such as the tank geometry) in the nonlinear model are estimated using the closed-loop operation data with the level controller.

Conclusions

In the inferential model fitting example, we investigate the hybrid approaches including piecewise linear and artificial neural net embedded into the simultaneous model fitting platform. The simultaneous model fitting by specifying constraints in a single optimization running reduces the extra effort to validate the model manually with a priori information and to manually sort out the appropriate segment of the historical operation data for each block. In the tank model example, the physics-based model is used to identify the unknown parameters using the available operation data. The identified physics-based model can then be used as a priori constraint in the simultaneous dynamic estimation problems in an empirical model fitting which is the most common type of advanced process control system. The benefit of this example is to improve the accuracy of the empirical linear model fitting and reduce the amount of time and effort by imposing the nonlinear constraint from the physics-based model as a priori information.

References

[1] Darby, M. L., & Nikolaou, M. (2012). MPC: Current practice and challenges. *Control Engineering Practice*, *20*(4), 328-342.

[2] Beal, Logan, et al. "GEKKO Optimization Suite." *Processes*6.8 (2018): 106.

COMPRESSOR SURGE MODELING AND CONTROL

Gregory K. McMillan, Emerson Automation Solutions, Chesterfield Missouri

Biography

Greg retired from Solutia Inc. as a Senior Fellow after 33 years specializing in improving loop performance. Greg contracted as a consultant in DeltaV R&D for 12 years and is presently a Principal Senior Software Engineer in Simulation R&D for Emerson Automation Solutions focusing on dynamic modeling using the Digital Twin. Greg is the author of 80 articles and papers and 20 books. Greg's most recent books are *Advances in Reactor Measurement and Control*, *Tuning and Control Loop Performance 4th Edition*, *Good Tuning: A Pocket Guide 4th Edition*, and *Process/Industrial Instruments and Controls Handbook 6th Edition*. Greg is an ISA Fellow and received the ISA "Kermit Fischer Environmental" Award for pH control in 1991, the Control Magazine "Engineer of the Year" Award for the Process Industry in 1994 and was inducted with Greg Shinskey into the "Process Automation Hall of Fame" in 2001. Greg was honored as one of InTech's 50 most influential innovators for advancing automation and control technologies and received the ISA Life Time Achievement Award in 2010.

Abstract

Axial and centrifugal compressor control is exceptionally challenging due to the extraordinary speed and severity of problems and the extreme consequences in terms of plant safety and performance. The fastest and most dangerous phenomenon is surge. An axial or centrifugal compressor in surge can reverse flow in 0.03 seconds going from a large positive flow to a large negative flow. Often, the negative flow is not measurable by flow meters leaving the actual situation to your imagination.

A high fidelity dynamic real time model has been developed and integrated into a Digital Twin to enable a greater understanding of the special requirements in terms of automation system speed of response and special algorithms. The model uses a momentum balance thoroughly researched and verified. The normally unseen compressor characteristic curve to the left of the surge point is used enabling the model to show the jumps in flow on the positive slopes of the characteristic curve. The unique shape and speed of surge flow oscillations is used in the Digital Twin to explore, develop, prototype, test, justify, deploy, commission and maintain the best surge control system. The Digital Twin also enables not only training of operators but education of process and automation engineers on the extreme dynamics and control system opportunities.

A simple but powerful technique is developed for detecting a potential surge and updating the surge curve. Control strategies are detailed that proactively prevent surge while maximizing compressor efficiency. Unnecessary surge valve openings are avoided and the surge controller setpoint is optimized to lie on the longitudinal axis of efficiency ellipses. Feedforward is used to deal with sudden disturbances from changes in production rates and shutdown of downstream users. Additionally, a method of optimizing the compressor discharge pressure based on user downstream valve positions is developed. All of these control system opportunities can be demonstrated with the actual control system connected to the model.

Compressor Surge Curve and Setpoint

A surge point is the point on the characteristic curve where the slope goes to zero. The characteristic curve is a plot of pressure rise versus suction flow for a given speed or inlet guide vane position. A volumetric suction flow is typically used because of advantages in terms of being more representative of the operating conditions for surge prevention. When a discharge flow is measured, temperature and pressure compensation are used to get it back to suction conditions. Operation at the point of a zero slope is extremely tenuous because a very slight deviation to the left results in positive feedback and a surge cycle. It is important that the operating point be kept sufficiently to the right on a part of the characteristic curve with a significant negative slope providing some self-regulation by negative feedback (e.g., increase in pressure causes an increase in flow and consequential decrease in pressure).

The surge curve is the intersection of surge points. For a centrifugal compressor, the surge curve is an exponential curve with an increasing slope for increases in flow demand. The ideal surge setpoint should be parallel to the surge curve. This provides a constant operating margin and aligns well with the long axis of the efficiency ellipses. The bias to the computation of the surge setpoint is minimized by making the measurements, surge valves, and PID controller fast enough to react for the worst case disturbance to prevent getting close to the surge curve. Coming off output limits and proportional relative to integral action is optimized.

For axial compressors, the surge curve bends over at high pressures. Also, the consequences of surge tend to be more disastrous because of the higher flows and the possibility of extreme acceleration. One very large compressor would hit a high speed shutdown in a fraction of a second potentially damaging the impeller that cost several millions dollars causing production losses of many more millions of dollars a day. A speed derivative was needed to detect the start of the acceleration and proactively stop the compressor.

Compressor Characteristic Curve and Surge Path

Often not shown is the compressor characteristic curve to the left of the surge point B where the slope becomes positive creating positive feedback causing the suction flow to jump to a negative flow point C on the compressor characteristic that has a negative slope. The operating point then walks along the negative slope portion of the characteristic curve due to negative feedback until it reaches point D the start of the positive slope and positive feedback at which point it jumps to point A, If the surge valve is not sufficiently open, the operating point walks back to point B starting the whole surge cycle over again. The precipitous drop in flow of the first jump can be as fast as 0.03 seconds. Each surge cycle tends to cause a loss in compressor efficiency that becomes noticeable after a total of 10 or more cycles. While successive cycles can be most damaging, the accumulation of cycles at different times is also a concern. The compressor manufacturer may have made tests to identify the surge point but may not have completely detailed the characteristic curves. We will see how surge points can be identified online and a first principle model can provide the characteristic curves to the left of the surge curve in sufficient detail to show the extreme oscillations during surge.

Flow and Pressure Oscillations during Surge

The flow oscillation is characterized by jumps from positive to negative flow and from positive to negative flow that correspond to the path on the positive slope of the characteristic curve creating positive feedback. The walk along the negative slope in the peaks and valleys take about 1 second leading to a total oscillation period of about 2 seconds for plant compressors. For laboratory and pilot plant compressors, the period is 20 times and 5 times faster. The fast periods for all compressors and incredibly fast jumps to very negative flows make recovery by just feedback control improbable.

The pressure oscillations have a period as fast as the flow oscillations but the jumps are slowed down and the amplitude attenuated by the discharge volume of the compressor. The best way of stopping the pressure oscillations is stopping the flow oscillations by an open loop backup that will hold the surge control valve open long enough for system to stabilize.

Compressor Surge Protection and Efficiency Maximization (Vent Valves)

A surge setpoint offset to the right of the surge curve on the compressor map is computed for a suction flow (surge) controller from the pressure rise across the compressor. The output of the PID surge controller goes to redundant vent control valves. The feedback controller is useful to prevent surge for slow approaches to the surge curve but needs help from control logic that is not operating in feedback (closed loop) mode to preemptively position and hold the surge valves open. Since this action is not closed loop and is meant to take over in potentially damaging situations, it is termed an open loop backup. When triggered, the open loop backup puts the PID surge controller in remote output putting the control valves in a position large enough to prevent surge.

The open loop backup must be able to detect surge as soon as possible. This is done by detecting the precipitous drop in flow that is the unmistakable indicator of the start of surge.

Since a surge cycle is extremely disruptive possibly causing the shutdown of downstream users and each surge cycle counts to a total that causes a noticeable loss in efficiency, it is extremely desirable to predict too close of an approach to the surge curve. This can also be done by the open loop backup. The key component for both recovery and prediction is a deadtime block to create a low noise reliable rate of change signal that can be updated as fast as the PID execution rate. For recovery from surge, a high rate of change triggers the backup that is the input minus the output of the deadtime block divided by the block deadtime that is typically just large enough to provide a good signal to noise ratio. For prevention of surge, a predicted flow one deadtime into the future that is too close to the surge curve triggers the open loop backup. The predicted flow is simply the input to a deadtime block plus the output with the block deadtime set equal to the total loop deadtime. Deadtime blocks can also be used to compute a pressure rate of change for a flow rate of change when operating to the right of the surge curve. A small pressure rate of change for a significant flow rate of change indicates an operating point close to the surge curve.

Turning on external-reset feedback (e.g., dynamic reset limit) enables the use of up and down rate limits on the setpoint the PID is manipulating that does not require retuning of the PID. External-reset feedback prevents the integral action from changing the PID output faster than what is happening in terms of the response to the PID output. This is called directional move suppression. Model Predictive Control greatly benefits from move suppression. By enabling a different move suppression depending on direction opens up many opportunities here and for many other applications. By putting up and down setpoint rate limits on the analog output block for the surge valves and fast readback of actual position, you can have a fast opening surge valve for surge recovery and prevention, and a slow closing valve for gradually returning to a more optimum operating position.

Directional move suppression is also useful for a valve position controller that seeks to lowers the compressor pressure to the point that keeps the furthest open downstream user valve at a good maximum throttle position. Directional move suppression by up and down setpoint rate limits on pressure controller will provide a slow approach to an optimum and a fast getaway for surge prevention. Not shown is a feedforward signal that would be added to the surge flow controller output to proactively open surge valves to a position that would balance a large decrease in flow by a downstream user. The feedforward signal would need to go through a signal characterizer that would compute the change in the X axis (change in valve signal) for a desired change in flow signal per the valves installed flow characteristic.

Compressor Surge Protection and Efficiency Maximization (Recirculation Valves)

The same kind of control strategies shown for vent valves can also be used for recirculation valves. There is a heat exchanger in the recirculation line to prevent heat buildup recirculating flow through the compressor. Recirculation valves are used for multi-stage compressors and compressors in series. It is desirable that stage or compressor have its own surge curve and control system. Decoupling signals can be used as necessary to prevent interaction. Compression rate dividing can be used to provide desired pressure rises for stages and compressors in series. If there are users of intermittent stage or compressor flows, feedforward signals would be applied to the respective stage or compressor surge controller.

Sources of Automation System Dynamics

It is critical that all the sources of delays and lags in a surge control loop be identified and minimized. The deadtime of the actual flow and pressure response (process response) is incredibly fast. The automation system is much slower and must be made as fast as possible to catch up with process response. The control valve often has the slowest 86% response time that is the summation of the pre-stroke deadtime and the rate limited exponential response (e.g., slewing rate with two lags). Signal filtering and transmitter damping must be minimized along with I/O scan times and filters and PID execution rates. Unsuspecting I/O filters are often the main limitation to the speed of response of a supposedly extra fast PID. The default values of these I/O filters may not be accordingly decreased to coincide with faster scan times offered. This mistake and many others are the result of not realizing all the potential contributors to the automation system deadtime that is the summation of all the delays plus equivalent deadtime from lags in series.

How to Make Valve Response Faster

Never replace a positioner with a booster despite age old rules that recommend this for fast loops. This can cause erratic dangerous movement (e.g., slamming shut) of a butterfly control valve from positive feedback due to high and low sensitivity of booster output and input, respectively.

A volume booster should be put on the positioner output with a bypass valve open just enough to prevent fast cycling by enabling the positioner to see some of the large volume of the actuator besides the small volume of the booster inlet port. The positioner should not have integral action and the backlash (deadband) and stiction (resolution) of the control valves must be minimized to less than 0.2% by the use of true throttling valves rather than tight shutoff on-off valve posing as throttling valves with deadband and resolution that can be order of magnitude or more worse (e.g., 2 to 10% deadband and resolution). The entire control valve assembly must be tested for small and large signal changes to verify the 86% response time is always faster than required (e.g., less than 1 second).

Valve Alerts

The most common mistake is selecting control valves based on tightest shutoff. These have the greatest seat or seal friction and consequently greatest stiction near the closed position. The next is maximizing capacity not realizing that any resolution limit or deadband is a percent of valve capacity and actual rangeability is lost preventing good control near the closed position. The other common mistake is seeking to minimize valve cost not getting the best most precise and responsive throttling valve.

There are many watch-outs. Actuators are often undersized based on thrust or torque requirement for loose packing or room temperature. The actuators should be sized to provide at least 150% of the thrust or torque required for maximum operating pressure and temperature. Resolution and deadband test should be done near the closed position where these effects are greatest (suppliers often choose 40% for testing valve response). The response time should be measured for small steps (e.g., 0.4%) and large steps (e.g., 40%) since actuator and positioner sensitivity and relay capacity come into play for small and large steps, respectively. On-off valves posing as control valves often have a shaft feedback that is not representative of ball or disk position due to friction in seal and packing and play in the actuator shaft to stem and from stem to actual ball or disk connections. A splined connection should be used for shaft to stem connection and the stem should be integrally cast with ball or disk (no connection). If this is not done, the smart positioner is being lied to by its position feedback giving diagnostics and readback that are not real leaving the user clueless.

Boosters should never be used instead of a positioner as previously detailed and integral action that makes the response to medium size steps look pretty, cause poor small and large valve response because the positioner gain needed is less than highest value to accommodate integral action. Also, integral action in the positioner will cause limit cycles from deadband when the surge controller is in automatic and cause limit cycles from stiction even when the surge controller is in manual.

Digital Twin Use and Improvements

The Digital Twin with a momentum balance to simulate the unique and very challenging requirements to prevent and recover from surge and automation system dynamics can be added to understand, design and verify compressor control systems maximize process safety, efficiency and capacity. The extreme consequences of less than the best control strategies and automation system components can be detailed and corrections can be made by rapid exploration and testing. The only limit is your imagination. Online metrics can show the benefits in terms of less loss of compressor efficiency and better process performance.

Digital Twin's Key New Blocks

Blocks have added to the Digital Twin to test innovations and improvements in strategies, dynamics and PID tuning. The Compressor W/Surge block uses a momentum balance to provide the dynamics of surge. The block can be put in series for multi-stage compressors or compressors in series. A Backlash-Stiction block can be put in series with each other to simulate what is happening in the actuator, shaft to stem connection, and stem to ball or disk connection. Blocks to include a variable deadtime, slewing rate and response lags can be inserted. The Future Value block can provide a timely low noise rate of change and future value typically a factor of the total loop deadtime into the future from current value. This is extremely useful for surge curve and actual surge identification.

Details on Control System Implementation and Best Practices

The principles and knowledge are listed as implementation details and best practices.

Conclusion

This is just a brief insight into the incredible deep understanding gained and great advances possible by compressor modeling and control.

References

1. McMillan, G.K., Stuart, C., and Hildebrand, T., "Compressor surge control: Deeper understanding, simulation can eliminate instabilities", *Control*, Oct 2018 https://www.controlglobal.com/articles/2018/compressor-surge-control-deeper-understanding-simulation-can-eliminate-instabilities/?stage=Live
2. McMillan, G.K. and Berutti, M.E., "Virtual plant virtuosity", *Control*, April 2017 https://www.controlglobal.com/articles/2017/virtual-plant-virtuosity/?stage=Live
3. McMillan, G.K. and Wu, P., "How to specify valves and positioners that don't compromise control", *Control*, May 2016 https://www.controlglobal.com/articles/2016/how-to-specify-valves-and-positioners-that-dont-compromise-control/?stage=Live
4. McMillan, G.K., *Centrifugal and Axial Compressor Control*, Momentum Press, 2010 http://www.momentumpress.net/books/centrifugal-and-axial-compressor-control

549081 A Model-Centric Framework for the Smart Manufacturing of Polymers

Plastic and polymeric materials have an enormous impact on a variety of industries and our daily lives. The synthesis of polymers occurs by reacting small molecules referred as monomers together to form, under certain conditions, long molecules referred as polymers. During polymerizations, it is mandatory to assure the uniformity between batches, high-quality of products, energetic efficiency, minimum impact to the environment, reliability of the process, and labor safety. It remains as a major challenge the establishment of operating, monitoring, and control conditions capable to achieve all mentioned goals together.

In this work, various model-centric strategies towards the smart manufacturing of polymers were designed, implemented, and tested using experimental data from two polymerization systems. The main aims included the combination of different model-centric strategies which permit to obtain operational conditions such that production goals are achieved. After validating the correctness and reliability of the fundamental model, different optimal policies are formulated, advance control elements are developed, and nonlinear online estimators are designed for reaching the joint goal of signal processing and full polymer characterization.

The first system studied is the synthesis of polyacrylamide in aqueous solution using potassium persulfate as initiator in semi-batch fashion. In this system, the proposed framework integrates nonlinear modelling, dynamic optimization, advanced control, and nonlinear state estimation towards producing polymers with target properties. The main objectives include the attainment of desired polymer characteristics through feedback control while a complete polymer motoring is possible during the reaction. Results, in terms of the dynamic set point achievement, are satisfactory for the closed-loop system. An improvement of 42% in average is foreseen when using a hybrid discrete-time extended Kalman filter (*h*-DEKF) as state observer. In addition, other strategies for state estimation are proposed including different structures of the geometric observer (GO) from which the 4-state GO with passivation shows the best performance overall. The 4-state GO with passivation achieves an improvement of 52% in the estimation of the molecular weight distribution when compared with the *h*-DEKF.

The second system is the copolymerization of ethylene with 1,9-decadiene using dimethylsilyl (N-tert-butylamido) (tetramethylcyclopentadienyl) titanium dichloride (CGC)/MAO as catalyst in semi-batch fashion. For this system, a data-driven strategy for online estimation of important kinetic parameters was assessed. A global sensitivity analysis was performed for selecting the most representative kinetic parameters of the system. The retrospective cost model refinement (RCMR) algorithm is adapted to the problem and implemented for online parameter estimation. After verifying the stability and robustness of the method, real experimental data permitted to validate its performance. Results show that the estimation of the kinetic parameters of interest is feasible, and converges close to theoretical values without requiring previous knowledge of their magnitudes.

548842 Jetty Bog Recovery

Jetty BOG Recovery

Yoshitsugi Kikkawa,

Kikkawa Giken, Yokohama Japan,

yoshikikkawa@yahoo.co.jp

David Hill

Manager of Technical Support, Chemstation, Houston,Texas

davidh@chemstations.com

Abstract

A considerable amount of BOG is generated during the LNG tanker loading for the base load LNG plant. Jetty BOG of the old generation baseload LNG has not been recovered, since the jetty BOG generation is intermittent and to recover the jetty BOG is costly. However recent concern about the global warming by CO2 emission from the flare stack, the jetty BOG recovery has become significant issue, nowadays. Generally speaking, the BOG is leaner compared with the baseload feed gas, therefore to recovery the BOG as feed gas will have some effect to the liquefaction performance. This study has checked the liquefaction performance when the BOG has been recovered as feed gas to the liquefaction process, using process simulation program CHEMCAD by Chemstation.

550309 Analysis of Process Dynamics during Transient Operations on LNG Vessel Gas Management System

Analysis of process dynamics during transient operations on LNG vessel gas management system

Nobutaka Umeyama Japan Marine United Corp.

Kouji Kikuchi, Kenichi Matsuoka Azbil Corp.

David Hill, Aaron Herrick CHEMSTATIONS Corp.

During LNG transportation on an LNG vessel, a significant fraction of the LNG cargo evaporates as boil-off gas (BOG). The rate of BOG generation fluctuates based on several factors, such as weather and tank contents. The dual-fuel engines (DFE) that power an LNG carrier have fluctuating fuel demands based on factors such as vessel displacement, the speed of the ship, and power demand. Additional BOG can be generated by vaporizing LNG. It is undesirable to flare or vent excess BOG, and most vessels are not equipped to re-liquefy BOG. The control system including compressor and surge control attempts to match the BOG generation with the DFE fuel demand to operate the vessel.

During the LNG vessel design phase, compressor systems have to be examined carefully, in order to avoid setbacks such as compressor surges or start-up difficulties due to undersized / oversized capacity. Then, potential problems during plant start-up, emergency shutdown and various turn down conditions are able to be identified and eliminated in advance.

We developed a dynamic simulation model for LNG vessel gas management systems including compressor surge control, and conducted an analysis of process dynamics during transient operations (e.g. start-up, shutdown, trip and change of fuel gas compositions).

The dynamic simulator was used to model the gas management system (GMS) – compressor system with surge control etc., to handle BOG, along with a detailed BOG model from the LNG cargo tank. It was then used during the design of process control, validation of operating procedure controllability, and factory acceptance test (FAT) to validate distributed control system (DCS) functionality.

In this project, the use of dynamic simulation system has been summarized as follows:

(1)Process Study & Design of Control concept

(2)Validation & verification of control functions in DCS:

(3)Tune-up of plant control parameters, by integration with DCS during FAT

In this paper, we compare the simulation result with the actual process behavior, and show the effectiveness and accuracy of dynamic simulation. In addition, the paper shows that the period of time for the tune-up of plant control during start-up on site could be reduced by investigating the dynamic behavior of the process, initially using the optimum control parameters during FAT.

550320 A CFD Study of LNG Boil-Off Consider Both Heat Leakage and Vibration Effect

Liquefied natural gas (LNG) is becoming an increasingly important clean energy source and is used and transported worldwide. However, its low boiling point makes LNG very easy to evaporate and generate boil-off gas (BOG). The BOG generation not only causes the loss of cold energy but also builds up the pressure of storage tank, which could cause the tank damage or even explosion. Thus, BOG should be minimized at any economic possibilities during its translations. Many studies have been focusing on the BOG generation caused by heat leakage. However, few works based on our literature survey have quantitatively studied the factor of mechanical vibration that can also contribute to BOG generation.

In this study, the BOG generation of an LNG truck caused by both factors of heat leakage and transport vibration has been quantitatively studied by coupling computational fluid dynamics (CFD) and process simulations together. The commercial CFD software, ANSYS Fluent V 18, has been employed to develop the multiphase mixture model for modeling the vibration effect caused by LNG truck transportation, where a user-defined function (UDF) has been programmed with the dynamic mesh algorithm. Energy generated by mechanical vibration were identified through CFD simulation and exported to process simulation, in which a single tank with pressure outlet was modeled to quantify the amount of BOG generated during transportation. This study for the first time addressed two major root causes for BOG generation, which could benefit deeper understanding of BOG generation mechanism and BOG minimization measures during LNG truck transportations.

550755 Optimal Operation for LNG Multiple Ship Tanks Commissioning and Uploading Via Rigorous Dynamic Simulations

Natural gas is fastest growing energy economy. Its production and demands are increasing rapidly. Natural gas is transported as liquid over long distances. In 2017, liquefied natural gas (LNG) trade reached 293 million tonnes per annum (MTPA), as per Shell LNG Outlook 2018 report. LNG imports grew by 29 MTPA which is 30% more than what was expected. Spot cargo deliveries increased by 17% in 2017 over 2016. Natural gas and LNG production rates are expected to increase over next couple of decades.

In this work, rigorous dynamic process simulation for commissioning and LNG loading of multiple LNG ship-tanks has been performed to study the optimal procedures when new LNG ship-tanks are put to service or old LNG ship-tanks come back to service after maintenance. At the beginning, the tanks contain humid air at ambient temperature. Before LNG can be loaded, the tanks require drying, inerting, gassing up, and cooling. Drying is necessary to avoid moisture from air to condense and solidify during cooling of the tank. Air is purged using inert gases to remove oxygen from the tanks. Then methane vapors are used to remove inert gases because some inert gases like CO_2 can solidify after adding LNG and can chock valves. Cold methane is used for initial cooling of the tank. LNG liquid is sprayed for further cooling to desired temperature. Finally, LNG is loaded into the ship tanks. The LNG ship-loading is constrained by maximum allowable cooling-rate of ship-tanks, jetty boil-off gas (BOG) compressor capacity, LNG loading line capacity, and LNG level in ship-tank. Overall, LNG loading is complicated process. Dynamic simulation is helpful to study the commissioning and loading operations. The changes in the composition of material inside the tank are reported for each step in the commissioning and loading procedures. Optimization of the process to minimize time of commissioning and loading operations, emissions, and energy requirements is carried out using the Aspen Plus Dynamics simulation tool. This study would help LNG industry to operate LNG exporting terminal smoothly and in an optimal way.

550940 Operations Improvement in LNG Re-Gasification Terminal through Corrected Storage Tank Model

The simulation of liquefied natural gas (LNG) storage tanks is often grounded on numerous problematic assumptions, for example, assessment of boil-off gas (BOG) generation using boil-off rate, static liquid-level, vapor liquid equilibrium in the tank, use of only lateral area for heat loss calculations, and etcetera. Few of these assumptions are built into frequently used commercial simulators, making further challenges in simulating the real behavior of LNG tanks. This study highlights these challenges in the background of a commonly used process simulator, Aspen HYSYS, and provides analytical and intuitive solutions to those problems. Then the corrected tank model is used for operations improving in LNG re-gasification terminal. Based on the study following rules of thumb for enhanced operations of LNG re-gasification terminal are deduced:

- The pressurization rate of the tank with the highest liquid level is the slowest
- Ambient heat leak dominates BOG generation at a low re-circulation rate
- In-tank pump work dominates BOG generation at a high re-circulation rate
- Minimum re-circulation that prevents 2-phase is the best operating policy

If implemented, the deduced rules could help enhance LNG terminals' energy efficiency and reduce operational costs while increasing safety.

550549 Monitoring Process Degradation through Operating Regime Based Process Monitoring

Process monitoring is an essential component of many industrial processes, and is required in order to maintain process safety, and minimize operational costs [1]. Process degradation through equipment deterioration can lead to a significant increase in operational costs, and thereby result in high economic losses.

Monitoring equipment deterioration can be a challenge as these parameters cannot be directly measured. Fortunately, state estimation methods are available, which can be utilized in order to estimate the values of these process parameters, and additionally other state variables, including those that cannot be easily measured using sensors. The family of Kalman filter estimation methods, including the Extended Kalman Filters (EKF) and Unscented Kalman Filters (UKF), have been utilized by many in order to carry out effective state and parameter estimation for different applications [2]. Many industrial processes, and pieces of equipment are known to follow nonlinear models. Recently, state estimation through particle filters has also received a reasonable amount of attention.

This work involves monitoring of a process using state estimation methods in order to identify the economic impact of equipment deterioration over time. Application of the algorithm will be demonstrated through a nonlinear example, e.g., a heat exchanger that experiences increasing levels of fouling over time. The presence of fouling causes the overall heat transfer coefficient to decrease, making the heat exchanger less efficient. This in turn forces the heat exchanger to utilize more cooling utility in order to maintain the same set points, thereby increasing operational costs. The example will estimate the parameter, i.e., overall heat transfer coefficient, and the states using different state estimation methods [3]. Additionally, dynamic data will be obtained by simulating the nonlinear heat exchanger in ASPEN Plus Dynamics, in order to enhance of the validity of the proposed algorithm.

References

[1] M.Z. Sheriff, F. Harrou, M. Nounou, Univariate process monitoring using multiscale Shewhart charts, in: 2014 Int. Conf. Control. Decis. Inf. Technol., IEEE, Metz, France, 2014: pp. 435–440.

[2] M.M. Mansouri, H.N. Nounou, M.N. Nounou, Nonlinear control and estimation in induction machine using state estimation techniques, Syst. Sci. Control Eng. 2 (2014) 642–654. doi:http://dx.doi.org/10.1080/21642583.2014.956842.

[3] M.S. Peters, K.D. Timmerhaus, Plant design and economics for chemical engineers, 4th ed., McGraw-Hill, 1991.

550677 Specifying Control Valve Performance to Achieve Desired Process Performance

In the 1990's, a study of over 5000 control loops revealed that 30% of the loops were INCREASING process variability due to poor valve performance. On a recent control foundation improvement project, over 50% of the loops had poor performance due to significant control valve performance issues. Poor control valve performance continues to be a widespread source of poor control loop performance. A control valve may be properly selected based on physical specifications, but its operational performance may not meet the requirements to achieve the required process performance. Properly specifying valve performance metrics such as dead band, resolution, process gain, step response time, and response regions are vital. Industry standard ISA/ANSI-75.25 provides definitions and test procedures for positioning and response time performance of control valve but does not provide guidance on selecting appropriate control valve performance specifications for a specific application. This presentation explains the impact of valve performance parameters and how to specify them based to achieve the desired process performance. The definition and importance of these performance metrics will be explained, and field examples will be provided. The metrics can also be used to diagnose and improve existing control valve performance when it does not meet the required control loop performance. The benefits include longer valve life, reduced process variability and improved process performance.

549828 Detection and Isolation of Abnormal Event in Nonlinear Industrial Processes By a Novel Data-Based Method

Detection and Isolation of Abnormal Event in Nonlinear Industrial Processes by a Novel Data-Based Method

Chiranjivi Botre[a], Hazem N. Nounou[b], Mohamed N. Nounou[c], and M. Nazmul Karim[a]

[a] *Artie McFerrin Dept. of Chemical Engineering, Texas A&M University, College Station, Texas 77843, USA*

[b] *Electrical and Computer Engineering Program, Texas A&M University at Qatar, Doha, QATAR,*

[c] *Chemical Engineering Department, Texas A&M University at Qatar, Doha, QATAR*

Abstract

Detection and isolation of abnormal events are important aspects in the industry to ensure safe and proper operation of the plant. It is important to accurately, reliably, and quickly detect changes or faults within the process, determine the cause of these changes, and apply corrective actions to mitigate risk. Multivariate statistical techniques are powerful tools that utilize data based model reduction methods capable of efficiently handling process noise and correlated data sets. In literature, multivariate statistical methods are widely discussed for applications in process monitoring and fault detection. Moreover, data based process monitoring techniques have been successfully applied to applications where the accurate process model are not available. Fault detection can be carried out in two phases; data based model reduction and statistical fault detection.

The partial least square (PLS) method is an input-output model that has been effectively applied to linear processes and is most commonly used for fault detection applications. A kernel extension of PLS has been proposed to provide an effective technique for modeling nonlinear industrial processes. In order to enhance the fault detection performance of KPLS model we have developed a new method that optimizes KPLS model through the use of a multi-objective genetic optimization approach. Our multi-scale kernel partial least square (MSKPLS) - based generalized likelihood ratio test (GLRT) method, has the ability to handle process noise, non-normal data distribution, and auto-correlated data sets to provide an effective fault detection technique that can be applied to nonlinear industrial process data.

Once process faults have been detected, accurate isolation of the specific fault within the plant must be achieved to aid swift corrective action. Our approach to fault detection using the advanced statistical GLRT method also allows for multiple faults in the process to be isolated. The isolation method is called a contribution plot and uses the calculated statistical values to determine specific process variables responsible for the process fault, narrowing the possible fault location from plant-wide consideration to the contributions of a key sub-block consisting of a few unit operations. We have developed a novel fault isolation and identification hybrid approach by using a targeted approach where the generalized likelihood ratio (GLRT) based contribution plot is used to isolate the faulty sub-block of a process, and then hybrid model-based observer with a neural network is used to isolate and identify the faulty equipment. We will demonstrate the improved fault detection and isolation algorithm using the nonlinear simulated continuously stirred tank reactor (CSTR) data and Tennessee Eastman Process problem as a case study.

Keywords: KPLS, Wavelet function, Hybrid observers, Fault isolation, Neural network, Continuously stirred tank reactor (CSTR).

References:

1. Bakshi, B. R. Multiscale PCA with application to multivariate statistical process monitoring. AIChE J. 44, 1596–1610 (1998).

2. Botre, Chiranjivi, Majdi Mansouri, Mohamed Nounou, Hazem Nounou, and M. Nazmul Karim. 2016. "Kernel PLS-Based GLRT Method for Fault Detection of Chemical Processes." Journal of Loss Prevention in the Process Industries 43 (September): 212–24.

3. Botre, C., Mansouri, M., Karim, M. N., Nounou, H. & Nounou, M. Multiscale PLS-based GLRT for fault detection of chemical processes. J. Loss Prev. Process Ind. 46, 143–153 (2017).

4. MacGregor, J.F., and T. Kourti. "Statistical Process Control of Multivariate Processes." Control Engineering Practice 3, no. 3 (March 1995): 403–14.

5. Teppola, Pekka, and Pentti Minkkinen. 2000. "Wavelet–PLS Regression Models for Both Exploratory Data Analysis and Process Monitoring." Journal of Chemometrics 14 (5-6): 383–99.

AUTHOR INDEX

AUTHOR INDEX

AIChE
120 Wall Street, FL 23
New York, NY 10005-4020

ISBN 978-1-5108-8702-2

CURRAN ASSOCIATES INC.
proceedings.com